Discovering Mathematics

Discovering Mathematics

The Art of Investigation

A. GARDINER

Department of Mathematics
University of Birmingham

CLARENDON PRESS • OXFORD

Oxford University Press, Great Clarendon Street, Oxford OX2 6DP

Oxford New York
Athens Auckland Bangkok Bogota Bombay Buenos Aires
Calcutta Cape Town Dar es Salaam Delhi Florence Hong Kong Istanbul
Karachi Kuala Lumpur Madras Madrid Melbourne Mexico City
Nairobi Paris Singapore Taipei Tokyo Toronto Warsaw

and associated companies in
Berlin Ibadan

Oxford is a trade mark of Oxford University Press

Published in the United States
by Oxford University Press Inc., New York

First published 1987
Reprinted 1998 (with corrections), 1989, 1991, 1992, 1994, 1998

British Library Cataloguing in Publication Data
Gardiner, A.
Discovering mathematics: the art of
investigation.
1. Mathematics—1961–
I. Title
510   QA37.2

Library of Congress Cataloging in Publication Data
Gardiner, A. (A. D.)
Discovering mathematics.
Bibliography: p.
Includes index.
1. Problem solving.   2. Mathematics—Problems,
exercises, etc. I. Title.
QA63.G37 1986   510   86–12759

ISBN 0 19 853265 2 (pbk)

Printed in Great Britain by
Biddles Ltd, Guildford and King's Lynn

Some [...] say that the introduction of heuristic style would require the rewriting of textbooks, and would make them so long that one could never read them to the end. [...] The answer to this pedestrian argument is: let us try.

(Imre Lakatos, *Proofs and Refutations*)

Preface

This book has been written to help readers investigate some challenging, but elementary, mathematics **by working on their own.** The two parts of the book form a natural progression: my whole aim in writing has been to persuade as many readers as possible to work through *at least one of the extended investigations in Part 2.* These extended investigations were originally conceived as a challenge for bright High School students—a challenge which would at the same time give them some insight into how one actually *does* mathematics. They have been used successfully with many groups of such students. But they have been particularly successful with undergraduates in Colleges and Universities. Time and again these students have responded with unexpected freshness, commitment, interest, and sheer imaginative inventiveness. The material has also been enjoyed by groups of teachers on in-service courses and by professional mathematicians.

The habit of investigation is rather poorly developed in most students. Their view of mathematics is essentially passive. The average student imagines that one must first be taught certain standard methods, and is then free to exercise one's own mind only to the extent of using these methods to solve standard problems—largely by imitation.

This book presents a very different view of mathematics. The reader should have little difficulty with the basic mathematical techniques which are needed: most will be familiar, and they are in any case entirely elementary. This should make it possible to concentrate on

    the way this elementary material is explored and developed,

and

    *the strategies we use to tackle simple-looking questions whose
    answers are not at all obvious at first sight.*

This emphasis on **the way one investigates a problem** should give a more realistic picture of the nature both of mathematics itself, and of mathematical thinking.

Part 1 contains a number of *short* investigations. These serve as a stepping-stone between ordinary 'puzzles'—in which a clearly stated problem is solved, and then abandoned—and the serious mathematics of Part 2, where one starts out from some interesting observation or question, and pursues it in several directions until some more or less satisfactory picture emerges, which we then try to explain. The experience of tackling one or more of the short investigations in Part 1 should help to build up the confidence and stamina of those who are not in a position to plunge straight into Part 2. The contents of Part 1 will also provide prospective and practising teachers with a source of problems for investigation in the classroom.†

Part 2 contains two detailed *extended* investigations, together with outlines for several other investigations of the same kind. Their mathematical content is elementary, but they are by no means easy. Each investigation is long—probably much longer than most readers are used to. (This is one of the reasons why Part 1 has been included.) But they have been written in such a way that eventual success depends mainly on persistence, and does not require any exceptional ability on the part of the reader. Those who complete one or more of these extended investigations will find that their length is fully justified and that the effort has been well worthwhile.

Investigation in mathematics is not an end in itself. It is a way of uncovering and of understanding significant mathematical patterns and relationships. A good textbook will make sure that whenever a new topic is introduced its most significant features are clearly visible and easily understood. But in real life the significant features of a problem are usually rather hard to identify, and a lot of painstaking detective work may be necessary before they can be finally tracked down. It is one of the most striking characteristics of mathematics that thoughtful and persistent mathematical analysis often throws up totally unexpected insights into what may at first have looked like an uninteresting, or intractible, problem.

The pace and style of the book are such that reasonable students should be able to work through the text and the exercises with the minimum of assistance. Many of the exercises have supplementary hints, and solutions to the exercises may be found at the end of each chapter. But the written word—whether text, exercises, hints, or

---

† My book *Mathematical Puzzling*, Oxford University Press 1987, contains many other challenging puzzles and investigations in a similar—but simpler—vein.

solutions—has its limitations. So the experience of working through this material should be supplemented by whatever student–student or student–teacher interaction can be arranged. The lists of 'Further problems' which follow each detailed investigation in the text provide a source of extra ideas for the teacher or lecturer who wants to use the material in the classroom, and give the general reader something to explore in the same spirit as the structured investigations which precede them.

I am indebted to many people for ideas, inspiration, and encouragement. (Ideas which one fondly imagines to be original usually turn out to have their roots elsewhere.) The material was conceived while visiting Deakin University and the University of Melbourne in 1981. Without Derek Holton's irrepressible enthusiasm the book might never have been born. Since then many friends and students have commented on successive drafts. I am grateful to them all—but especially to Tom Beldon, whose comments and 'innocent' questions have challenged me all the way.

Birmingham   A. G.
July 1985

Contents

Introduction

Where there are problems, there is life.
(A. A. Zinoviev, *The Radiant Future*).

This book has been written to help you investigate some challenging, but elementary, mathematics **by working on your own.** Part 1 contains four short investigations, in which the challenge is for you to explore and explain what is going on rather than just to find the right answer. Part 2 contains two extended investigations. These are more demanding than the short investigations in Part 1, but their mathematical content is still entirely elementary. Where you choose to begin, and just how much time you spend on each part, will depend on your previous mathematical experience. But my aim is to persuade you to eventually work through at least one of the extended investigations in Part 2.

In each part certain mathematical 'problems' have been singled out for special treatment. Each such problem is introduced, and then explored, in the text via a structured sequence of exercises. Each of these structured investigations is followed by a list of 'Further problems' of roughly the same kind for you to have a go at on your own.

The material makes extensive use of basic techniques from secondary school mathematics. But the particular problems studied, and the way in which they are studied, lie outside most mathematics syllabuses. These problems have their own intrinsic interest, but they are not in themselves all that important. What is important is *the way the problems are studied*. In particular, I have tried to present the material in a way which will help you:

(1) to develop *strategies* for solving mathematical problems,

(2) to practise *the art of guessing* intelligently *and then testing* your guesses,

(3) to appreciate *the need for proof*,

(4) to see how *thoughtful generalization* can give rise to *a reformulation of the original problem*, which may then be easier to solve,

(5) whenever possible to *apply hard-won results and methods* to answer related questions,

(6) to appreciate the need for *discipline and stamina* when doing mathematics.

So, though the problems have been carefully chosen, they are chiefly a means to an end—and that end is

to cultivate the art of *doing* mathematics.

Some of the short investigations in Part 1 result from an attempt to solve a clearly stated puzzle or problem. Others begin with an experimental investigation of some curious mathematical phenomenon. In each case you should explore the problem intelligently until you not only know *what* is going on, but can also explain *why*.

The two investigations in Part 2 may be tackled in either order. The second investigation has the advantage that the main conjecture can be formulated on the basis of very simple numerical work; but this does not get one very far, and other ideas soon have to be introduced. The first investigation does not begin in quite such a simple-minded way. But it has the advantage that the initial naive approach can be sustained for much longer; and though it too has to be supplemented eventually by more powerful methods, it is never actually supplanted by them.

Part 1

Short investigations

Advice to the reader

. . . if you wish to learn swimming you have to go into the water.
(G. Polya, *Mathematical Discovery* Vol. 1)

Each investigation starts out with a **problem** of some kind. (The first and third investigations result from an attempt to solve a clearly stated puzzle or problem; the second begins with an 'experimental' investigation of a curious arithmetical phenomenon; the fourth is an attempt to generalize a familiar formula from two-dimensional geometry to three-dimensions.) The initial problem is then explored via a structured sequence of **exercises**, interspersed with **text**. In each case my aim is to get you to explore intelligently until you feel that you not only know *what* is going on, but can also *explain why*.

The exercises are like mathematical *bricks*. They provide the basic experiences you need to help you uncover some of the interesting mathematics which lies behind the original problem. The text is like the *mortar* which cements these exercises together. Some of the exercises have hints: these are printed at the end of the relevant investigation and before the list of Further problems. Solutions to the exercises are printed at the end of each chapter.

The text and exercises map out *one way* of approaching the initial problem. They are offered as a guide, not a straitjacket. I shall have succeeded only if you feel free to use the text as a starting point, and begin to ask *your own* questions and to explore the initial problem in *your own* way.

1 Analysing games

There is no possible escape from the situation.
(General Gordon's Khartoum Journal)

1.1 Introduction

Thinking about games, and how to win them, can be a rich source of mathematics. But thought requires effort, and games are meant to be 'fun'. So it is tempting to forget that effort can *be* fun, and to think about such games only in the most superficial way. (Many people, for example, play card games without even trying to remember which cards have been played!)

The main purpose of this chapter is to get you to explore one family of simple games for *two* players. The games themselves could equally well be played by three or more players, though our analysis of how one should play to win would no longer apply. But before you launch into the investigation I shall try to put these games and the way we analyse them into some kind of general context.

Lots of games are based on the following simple idea. One starts with a procedure which is completely predictable when carried out by a single person. One then introduces a subtle twist by requiring alternate steps to be carried out by two *competing* players. Instead of cooperating to complete the procedure as efficiently as possible, each player now tries to steal a march on the other. The strategy each player adopts will depend, for example, on

(a) whether the player who makes the final move is to be the winner or the loser, and

(b) whether the two players' moves can be told apart (as in 'Noughts and crosses'), or not (as in 'The efficient ruler game'—see Section 1.3, Further problem 9).

With simple games of this type it is always worth considering such variations as these to see which produces the most interesting play. You should bear this in mind when you come to analyse the following examples.

(1) *Noughts and crosses* The aim of getting three entries in a row would present no problems if either the two players did not

compete, or if their moves could not be told apart ('Noughts and noughts'): in either case, the first player could scarcely help winning.

(2) *The game of Euclid* (See Section 1.3, Further problem 7) The art here lies in choosing the right multiple of the smaller number to subtract each time. The game is an excellent example of competition shedding new light on an apparently routine algorithm. The winning strategies are unexpected, and delightfully simple.

(3) *The fifteen game* (See Section 1.3, Further problem 8) This is one of those games one would like everyone to meet. It may look pretty ordinary—but don't be deceived!

(4) *The efficient ruler game* (See Section 1.3, Further problem 9) Given a blank piece of wood of length 15 cm (say), it is an interesting problem to find the minimum number of marks you need to make in order to be able to measure each whole number of centimetres from 1 cm to 15 cm *directly*. The visual scanning required simply to check whether a given set of marks suffices, is itself a valuable mental exercise (see Fig. 1.1). But though the underlying procedure here is far from routine, the competitive twist of requiring alternate marks to be made by two competing players still has interesting consequences. Both players must then look beyond the immediate effect of each move, and consider the possibilities each new mark opens up for their opponent.

There are several different 'levels' on which one can analyse such games mathematically. I shall, for simplicity, distinguish only three such levels. These may be briefly described as

Level 1: *Local* reasoning
Level 2: The search for *global* rules
Level 3: Being absolutely sure.

All three levels are important. But from a mathematical point of view our aim must always be to reach Level 3 if we possibly can.

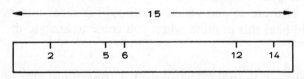

Fig. 1.1 An efficient ruler?

Level 1 Each time we make a move we have to ask ourselves what the *immediate* consequences of that move are likely to be: 'if I go there, then s/he'll . . .'. This kind of reasoning is *local* in the sense that it can only be used on *one little bit* of the game at a time. Such reasoning is important, but it ignores long-term effects. A move may be 'locally' safe, yet guarantee defeat in the long run.

Level 2 *Global* rules or strategies are those which influence the way one plays the game as a whole. These may be no more than 'rules of thumb' at first—like the 'Noughts and crosses' strategy of trying to set up a single move which creates *two* almost completed rows simultaneously. However, a fully-fledged global rule should be more dependable than a mere rule of thumb. It will nevertheless often be based on experience, observation, and guesswork rather than proof.

Level 3 Many people would be quite content to discover a global strategy which seems to work. But there are two reasons why it is worth trying to do better. First, one's initial guess often fails to tell the whole story, and it is only by pursuing the question of how to control play that one discovers the whole truth. (Examples (2) and (3) above are good examples of this.) Second, from a mathematical viewpoint it is not enough just to 'guess'—no matter how much evidence there may be that one's guess is correct. What one needs is some kind of mathematical *proof* that one's strategy really does control play in the way one thinks it does.

1.2 First short investigation

It is now time for you to do some work—to stop merely reading, and to start *thinking*! The particular game we shall investigate is meant to be fairly straightforward. It introduces several ideas which will be useful later on when you come to tackle much harder problems.

Problem Two players *A* and *B* have a pack of 9 cards. They take turns to remove 1, 2, or 3 cards from the pack—but a player must never remove the same number of cards as the previous player. The winner is the one who either takes the last card or leaves the other player with no valid move. Who wins?

Either the pack goes on getting smaller and smaller until someone takes the last card, or there comes a point where one player has no valid move. In either case someone has to win. But who?

The first player *A* can choose one of three moves. Can he perhaps always make sure of winning by choosing this first move carefully? Or can the second player *B* somehow regain control by always choosing the 'correct' response?

Exercise 1 Play the game a few times (if necessary against an imaginary opponent) and try to see what happens.

Exercise 2 Player *A* can make sure of winning only if he can remove just the right number of cards to leave Player *B* in a hopeless position.

(i) Suppose Player *A* starts by removing 3 cards. What should Player *B* do to make sure of winning?

(ii) Suppose Player *A* starts by removing 2 cards. Can Player *B* still make sure of winning?

(iii) Suppose Player *A* starts by removing just 1 card. Can Player *B* make sure of winning this time?

The original problem uses a pack of nine cards. This has the advantage that the resulting game is fairly easy to analyse. But if you wanted to challenge your friends it might be more interesting to start with a slightly larger pack. What would happen if one played with a pack of 21 cards? Or an ordinary pack of 52 cards? Or even a pack of 999 cards? Could the first player still somehow make sure of winning every time?

For a more interesting game one needs to start with more than nine cards, but not so many that each game takes too long. However this is not the only reason why it is worth asking what happens when we change the size of the pack. In the original problem with a pack of nine cards we observed that the first player can be sure of winning only if he can remove just the right number of cards to leave the second player in a 'hopeless' position. At first Player *A* will probably not know exactly how many cards to take. But no matter how many cards he removes, he will always leave Player *B* in the position of *playing first in a game with fewer cards*. So to choose his best move Player *A* must make sure that he leaves Player *B* in the position of playing first (with a smaller pack) in a game which he cannot possibly win! So if Player *A* is going to succeed in the game with nine cards, it looks as if he is going to have to know all about what happens in games where one starts with a smaller pack.

Exercise 3 (i) Can the player who plays first make sure of winning in the game with 6 cards?

(ii) Can the player who plays first make sure of winning in the game with 7 cards?

(iii) Can the player who plays first make sure of winning in the game with 8 cards?

(iv) What would you choose as your first move if you were playing first in the game with 9 cards? Why?

You may have noticed that we have apparently overlooked one small complication. If the first player A removes just one card, then Player B is not quite in the position of 'playing first in the game with 8 cards', since he is not allowed to remove the same number of cards as the previous player.

Exercise 4 (i) Which player *loses* the game with 8 cards? Will the same player lose if the first player is only allowed to take 2 or 3 (rather than 1, 2, or 3) cards on his first move?

(ii) Which player *wins* the game with 7 cards? Can the same player win if the first player is only allowed to take 1 or 3 (rather than 1, 2, or 3) cards on his first move?

(iii) Which player *wins* the game with 6 cards? Can the same player win if the first player is only allowed to take 1 or 2 (rather than 1, 2, or 3) cards on his first move?

Even if the game begins with a very large pack, as Player A's turn comes round again and again he finds himself playing first in games with fewer and fewer cards each time. So it is still important for A to know what happens when the starting pack is *small*.

Exercise 5 (i) Suppose Player A always starts. If the pack has just one card, Player A will remove it and win. Decide for yourself what happens if the starting pack has 2 cards, 3 cards, and so on up to 13 cards. Enter your results in the table in Fig. 1.2.

Number of cards in pack	1	2	3	4	5	6	7	8	9	10	11	12	13
Who can force a win?	A												

Fig. 1.2.

(ii) Who would you expect to win when the pack starts with 15 cards? What should their first move be? And why would this guarantee victory?

(iii) Who would you expect to win when the pack starts with 16 cards? What should their first move be?

(iv) Who would you expect to win when the pack starts with 52 cards? What if the pack starts with 999 cards?

Now go back and practise playing the game with different sized packs until you are sure that you know how to choose the best move each time. Then challenge a friend! You may occasionally have to start playing from a 'losing' position. (For example, if the size of the pack is fixed, then you will probably have to toss to decide who starts; or you may toss and allow the one who wins to choose the size of the pack, while the one who loses chooses whether to go first or second.) But even if you start from what should be a 'losing' position, you should usually be able to regain control by making the right moves—as long as your opponent does not discover how to win.

1.3 Further problems

1. Two players take turns to remove either 1 or 3 cards from a starting pack. The winner is the one who removes the last card. Who can force a win? Explain why (and how).

2. (i) Two players take turns to remove 1, 2, or 4 cards from a starting pack. The winner is the one who removes the last card. Who can force a win? Explain why (and how).

(ii) What happens if the players are allowed to remove 1, 2, or 5 cards (rather than 1, 2, or 4)?

(iii) What happens if the players are allowed to remove 1, 2, or 6 cards?

3. (i) Two players take turns to remove 1, 3, or 4 cards from a starting pack. The winner is the one who removes the last card. Who can force a win? Explain why (and how).

(ii) Who can force a win if neither player is allowed to remove the same number of cards as the previous player?

4. Two players take turns to remove 1, 2, or 3 cards from a starting pack, but the number of cards removed by the two players in two successive moves must never add up to exactly 4. (So if a player removes 3 cards, the next player is not allowed to remove just 1 card.) The winner is the player who either removes the last card or leaves the other player with no valid move. Who can force a win? Explain why (and how).

5. *The game of Nim.* The game starts with several piles of matches. Two players take turns to remove as many matches as they like from any one pile—and from one pile only each turn. The winner

is the one who takes the last match. When can the first player make sure of winning? And how should he play to win?

6. *Wythoff's game.* The game starts with two piles of matches. Two players take turns to remove either any (positive) number of matches from any one pile, or equal numbers of matches from both piles. The winner is the one who takes the last match. When can the first player force a win? How should he play in order to win?

7. *The game of Euclid.* Each player chooses a positive whole number and records it secretly. The two players then toss to decide who should start before revealing their chosen numbers—say a and b ($a \leqslant b$). The first player then changes the pair a, b by subtracting any multiple of the smaller number from the larger to produce a new pair a', b'. Negative numbers are forbidden. The second player can then transform this new pair a', b' in the same way, and so on. The first player to produce a pair in which one of the two numbers is zero is the winner. When can the first player force a win? How should he play in order to win?

8. *The fifteen game.* One starts with a row of nine squares labelled 1 to 9. Two players take turns to annex one vacant square per go. The winner is the first player to occupy three squares adding up to fifteen. Who can force a win? And how should he play to win?

9. *The efficient ruler game.* Two players start with an unmarked piece of wood 15 cm long (say). They take turns to make a mark at one of the fourteen points which measure a whole number of centimetres. The winner is the first player to produce a 'ruler' with which one can measure *directly* each whole number of centimetres from 1 cm up to 15 cm. Who wins?

Hints and references for the further problems

Hints

1–3. See Ref. 8, pages 131–6.

4. See Ref. 8, pages 264–6.

5. If there is only *one* pile, the first player wins immediately. It might help to start by exploring the game with just *two* piles. See Ref. 1, pages 42–4; Ref. 7, pages 36–8; Ref. 8, pages 137–54; Ref. 9, pages 20 and 103–5.

6. See Ref. 1, pages 62 and 76–7; Ref. 5, pages 101–5; Ref. 6, pages 46–9; Ref. 7, page 39; Ref. 9, pages 20 and 105–12.

7. See Ref. 2, pages 41–2 and 47–8; Ref. 6, pages 1–4.

8. Where else have you seen the numbers 1 to 9 adding up in threes to give 15? See Ref. 3, page 116.

9. It may help to start by analysing the game with a 2 cm long piece of wood, then the game with a 3 cm long piece of wood, and so on. The point

here is not that some obvious pattern is bound to emerge! But these simpler problems will give you the chance to *develop strategies* which might help you analyse the game with longer rulers. See Ref. 4, Chapter 15.

References

1. E. R. Berlekamp, J. H. Conway, and R. K. Guy *Winning ways* Vol. 1, Academic Press, London, 1982.
2. A. Gardiner *Infinite processes,* Springer, New York, 1982.
3. M. Gardner *Aha! insight,* Scientific American, New York, 1978.
4. M. Gardner *Wheels, life, and other mathematical amusements,* W. H. Freeman, New York, 1983.
5. R. Honsberger *Ingenuity in mathematics,* Mathematical Association of America, Washington, 1970.
6. J. Roberts *Elementary number theory,* MIT Press, Cambridge, MA, 1977.
7. W. W. Rouse-Ball and H. S. M. Coxeter *Mathematical recreations and esays,* University of Toronto Press, Toronto, 1974.
8. F. Schuh *The master book of mathematical recreations,* Dover, New York, 1968.
9. A. M. Yaglom and I. M. Yaglom, *Challenging mathematical problems with elementary solutions* Vol. 2, Holden-Day, San Francisco, 1967.

1.4 Solutions to exercises in 1.2

Exercise 2 (i) B cannot remove 3 cards. If B removes 2 cards, A must then remove 1 or 3 cards and B can then win. If B removes 1 card, A must remove 2 or 3 cards and B still wins.

(ii) If B removes 1 card, then A is left with a pack of 6 cards, and so can make use of B's strategy in (i) to win (by removing 2 cards). So to win B must remove 3 cards, leaving A with a pack of 4 cards. A then loses. (Why?).

(iii) If B removes 2 cards, then A is left with a pack of 6 cards, and so can make use of B's strategy in (i) to win (by removing 1 card). So if he wants to win, B had better try removing 3 cards: A will then remove 1 card (why?) and B will lose.

Exercise 3 (i) Yes—this is exactly what B did in Exercise 2(i).

(ii) Yes—this is exactly what B did in Exercise 2(ii).

(iii) Exercise 2(iii) suggests the answer 'No'. But to be sure you must check what happens if the first player removes 1 card: the second player will then be left with a pack of 7 cards and can make use of B's strategy in Exercise 2(ii) (by removing 3 cards).

(iv) Remove just 1 card. Exercise 2(iii) then shows that the second player loses.

Exercise 4 (i) The first player. Yes.

(ii) The first player. Yes (this is just Exercise 2(ii)).

(iii) The first player. Yes (this is just Exercise 2(i)).

Exercise 5 (i)

Number of cards in pack	1	2	3	4	5	6	7	8	9	10	11	12	13
Winner	A	A	A	B	A	A	A	B	A	A	A	B	A

(ii) *A*. Remove 3 cards: *B* is then playing first in the game with 12 cards (with the extra handicap of not being allowed to remove 3 cards on the first move), and the second player (*A*) can always win the game with 12 cards.

(iii) *B*. Always try to leave *A* with a pack whose size is a multiple of 4. If *A* tries to prevent this by removing 2 cards first go, *B* removes exactly 1 card (*A* must then remove either 2 or 3 cards and *B* can respond with 3 or 2 respectively—leaving *A* with a pack of 8 cards).

(iv) *B* (apply the same strategy as in (iii)). *A* (remove 3 cards to leave *B* with a pack of $996 = 4 \times 249$ cards).

2 From rhyme to reason

In mathematics there is no ignorabimus.
(David Hilbert)

2.1 Introduction

All good mathematics involves some element of surprise. At its simplest this may take the form of an unexpectedly persistent pattern. Some patterns fairly hit you in the eye. Others are far less obvious, and may not strike you until you have grappled with a problem for quite a while. And sometimes you may find no pattern at all—even though you have good reason to expect one. (You may find that the first of the Further problems in Section 2.3 involves patterns of all three kinds.)

But there is more to mathematics than spotting simple patterns. Students can be quite good at spotting such patterns, yet never learn that the most interesting mathematics arises from tracking down *more elusive* patterns and from trying to explain *why* they occur. It is part of the beauty of Mathematics that it is possible not only to recognize 'rhymes', but also to figure out the 'reasons' behind them.

The Short investigation in this chapter and the Further problems which follow it all happen to involve numbers, but that is incidental. What matters is that they are all *curious phenomena which demand explanation.* It should not take you too long to sort out *what* happens each time. The real challenge lies in trying to discover *how*, or *why*, it happens.

2.2 Second short investigation

Problem (i) Pick a three-digit number whose first and last digits are different.

a b c

Reverse the order of the digits to get another three-digit number.

c b a

Subtract the smaller of these two numbers from the larger one to get a new three-digit number. (If you get a two-digit number *ef*, write it as a three-digit number '0*ef*' with the first digit zero.)

$$
\begin{array}{r}
a\ b\ c \\
-\ c\ b\ a \\
\hline
d\ e\ f
\end{array}
$$

Reverse the order of the digits of your new three-digit number. (If your new three-digit number '0*ef*' has first digit zero, reversing it will give a three-digit number '*fe*0' with last digit zero.)

$$f\ e\ d$$

Finally add your new three-digit number '*def*' and its reverse '*fed*'. *What answer do you get?*

$$
\begin{array}{r}
+\ \ d\ e\ f \\
f\ e\ d \\
\hline
\bullet\ \bullet\ \bullet\ \bullet
\end{array}
$$

(ii) Pick another three-digit number '*abc*' and repeat the same process. What answer do you get this time?

(iii) Get a friend to pick a secret three-digit number. Give the same sequence of instructions you followed in part (i), but make sure they keep the answer secret. Then pretend to do some private calculations before announcing the answer.

(iv) Now try to figure out why you always get the same answer 1089.

When you meet a puzzle like this one, you should not be content just to treat the answer as a 'magic number'. Instead you should always try to figure out *why it occurs*. We shall come back to this in Exercise 3 below. (In Part 2 you will discover that there is more to this particular 'magic number' than meets the eye.)

One way of trying to make sense of an unexpected result like that in the problem above is to try to put it into some mathematical context by looking at closely related questions. We start by looking at two different questions. Exercises 1 and 2 consider what happens in other bases. Exercise 4 considers what happens if we start with a two-digit or a four-digit number instead of a three-digit number. Exercise 3 challenges you to explain what you observed in the above problem and what you discover in Exercises 1 and 2.

You may not have had much practice at doing arithmetic in other bases so it may be worth starting with one or two general remarks about bases. The underlying idea is very simple. For example

$$\text{two hundred and six} = 2 \times 10^2 + 0 \times 10 + 6$$

so we write for short

$$\text{two hundred and six} = \mathbf{2\ 0\ 6}_{\text{base 10}}.$$

But it is just as true to say that

$$\text{two hundred and six} = 2 \times 9^2 + 4 \times 9 + 8$$
$$= \mathbf{2\ 4\ 8}_{\text{base 9}}.$$

When doing addition or subtraction in base 9 you simply imitate what you do in base 10: the only difference is that in base 9 you 'carry' multiples of nine instead of multiples of ten: e.g. $8 + 3 = 12_{\text{base 9}}$, and $27_{\text{base 9}} + 63_{\text{base 9}} = 101_{\text{base 9}}$.

Exercise 1 This is a straightforward imitation of the Problem above, but this time all numbers are written in base 9.
   (i) Pick a three-digit number (base 9).

$$\mathbf{2\ 4\ 8}_{\text{base 9}}$$

Reverse the order of the digits.

$$\mathbf{8\ 4\ 2}_{\text{base 9}}$$

Subtract the smaller number from the larger number.

$$
\begin{array}{r}
8\ 4\ 2 \\
-\ 2\ 4\ 8 \\
\hline
5\ 8\ 3_{\text{base 9}}
\end{array}
$$

Reverse the order of the digits in the answer and add. *What answer do you get?*

$$
\begin{array}{r}
5\ 8\ 3 \\
+\ 3\ 8\ 5 \\
\hline
_{\text{base 9}}
\end{array}
$$

   (ii) How does your answer in base 9 fit in with the answer you got in base 10?

Exercise 2 What answer would you expect to get in base 7? Repeat Exercise 1—this time using a three-digit number (base 7). Does the answer you get agree with the answer you expected?

Even if you had difficulty explaining why the number 1089 kept cropping up in the original Problem you should now feel that an explanation is long overdue. And if you did manage to give a satisfactory answer to Part (iv) of the original Problem, you should now revise that answer to take account of what you found in Exercises 1 and 2.

Exercise 3 (i) Is the number 1089 the only possible answer to the original Problem (base 10)? If, so explain why. If not, give a complete list of all possible answers.

(ii) Write out a 'base 9 version' of your answer to Part (i). Does this explain what you found in Exercise 1?

(iii) Try to write out a 'base b version' of your answer to Part (i) which will show what happens in base b for any value of b.

You may have had difficulty giving the kind of explanation, or proof, that was required in Exercise 3. Whenever this happens it is worth looking for a simpler problem of the same type to practise on.

Exercise 4 (i) Repeat the original Problem in base 10, but this time starting with a two-digit number 'ab'. What answer do you get? Do you get the same answer no matter what two-digit number you start with? If so, explain why. If not, give a complete list of all possible answers and explain why it is complete.

(ii) If in Exercise 3(i) you had difficulty *explaining why* the number 1089 is the only possible answer, try to modify the explanation you gave for the two-digit problem above to improve your original answer to Exercise 3(i).

Even if you had no difficulty at all with Exercise 3, there remains the question of why the original Problem started with a *three*-digit number rather than a two-digit, a four-digit, or even a twenty-one digit number. You may feel that the two-digit problem would be less effective as a trick simply because the number 99 which crops up in Exercise 4 looks a bit too special. Still before finally abandoning the original Problem it may be worth investigating what happens when you start with a four-digit number.

Example 5 Repeat the original Problem in base 10, but this time starting with a four-digit number '$abcd$'. What answer do you get? Repeat this several times, starting with a different number each time.

Do you always get the same answer? Obtain a complete list of all possible answers and try to show that your list is complete.

You will see some of these numbers again in Part 2.

2.3 Further problems

1. Fig. 2.1 shows a variation on Pascal's triangle. Each entry is equal to the *difference* of the two numbers immediately above it. What interesting features does this number triangle have? How many of them can you explain?

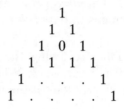

Fig. 2.1

2. Here is a simple procedure. What does it do?

Step 1   Choose a four-digit number (*say 7380*).

Step 2   Rearrange the four digits first to form the largest possible number (*8730*), then to form the smallest possible number (*0378*).

Step 3   Subtract the smaller number from the larger number and so obtain a new four-digit number (*8730 − 0378 = 8352*).

Step 4   Repeat Step 2 and Step 3 with your new four-digit number (*8532 − 2358 = 6164*).

Continue this procedure until something interesting happens. Then try another four-digit starting number. Try to decide what happens in general *and how long it takes*. Then try to *prove* what you believe to be true.

3. There is a well-known piece of 'number-magic' which goes like this.

MAGICIAN. Choose any number less than a thousand, but keep it to yourself.

VICTIM. (Thinks: 72.)

MAGICIAN. All I want you to tell me are the three remainders r, s, and t which you get when you divide your number by 7, 11, and 13. I will then tell you your original number.

VICTIM. $r = 2$, $s = 6$, $t = 7$.

MAGICIAN. (Thinks: $715r + 364s + 924t = 715 \times 2 + 364 \times 6 + 924 \times 7 = 10082$, subtracting as large a multiple of $7 \times 11 \times 13 = 1001$ as I can gives $10082 - 10 \times 1001 = 72$.) Your number was 72.

It is not clear why the trick appears in exactly this form. The divisors 7, 11, 13 are conveniently small, but the 'decoding' numbers $a = 715$, $b = 364$, and $c = 924$ which you have to use to recover the original number are unpleasantly large. There are therefore two good reasons why it is worth looking at simpler versions of the same trick. First, the principle on which it is based is a very general one, and you may find it easier to see what is going on in a simpler example. Second, the trick is most effective if you can do all the calculations with the decoding numbers *in your head*. Parts (i) and (ii) below present two simple variations of the original trick. In Part (i) the three decoding numbers a, b, c are given. Try to discover how they were chosen. Then use this to find your own decoding numbers for Part (ii). Practise both tricks until you can do the necessary calculations in your head every time. Then try them out on your friends.

(i) MAGICIAN. Choose a number less than a hundred, but keep it to yourself.

VICTIM. (Thinks: 73.)

MAGICIAN. All I want you to tell me are the three remainders r, s, t which you get when you divide your number by 2, by 3, and by 17. I will then tell you your original number.

VICTIM. $r = 1$, $s = 1$, $t = 5$.

MAGICIAN. (Thinks: If I choose the three decoding numbers $a = 51$, $b = 34$, $c = 18$, then no matter what r, s, and t are, the number $N = ar + bs + ct$ is bound to leave the remainders r, s, and t when one divides by 2, 3, and 17. If $r = 1$, $s = 1$, $t = 5$, then $N = 51 \times 1 + 34 \times 1 + 18 \times 5 = 175$. subtracting $2 \times 3 \times 17 = 102$ as many times as possible from this leaves $175 - 1 \times 102 = 73$.) Your number was 73.

(ii) MAGICIAN. Choose any number less than a hundred, but keep it to yourself.

VICTIM. (Thinks: 74.)

MAGICIAN. All I want you to tell me are the three remainders r, s, t which you get when you divide your number by 3, by 5, and by 7. I will then tell you your original number.

VICTIM. $r = 2$, $s = 4$, $t = 4$.

MAGICIAN. (Thinks: I want three decoding numbers a, b, c so that the number $N = ar + bs + ct$ always leaves remainders r, s, and t when you divide by 3, 5, and 7 respectively *no matter what values r, s, and t may happen to have*. Mmm, let me see . . . . Right got it! So in this case we have $N = 2a + 4b + 4c = 284$. Subtracting $3 \times 5 \times 7 = 105$ as many times as possible from the answer leaves $284 - 2 \times 105 = 74$.) Your number was 74.

4. (i) Choose any three-digit number—say 327. Its first and second digits differ by 1, its second and third digits differ by 5, its third and first digits differ by 4. In this way you get a new three-digit number 154. Repeating this process over and over again produces a sequence

$$327 \rightarrow 154 \rightarrow 413 \rightarrow 321 \rightarrow 112 \rightarrow 011 \rightarrow 101 \rightarrow 110 \rightarrow 011 \rightarrow \ldots.$$

At this stage you may as well stop. Why?

(ii) Repeat the whole procedure starting with the number 508. Then try lots of other three-digit numbers. What do you find? Can you explain why?

(iii) Try out the same procedure starting with some two-digit numbers. What do you find? Can you explain? Does your explanation work for three-digit numbers too?

(iv) What about four-digit numbers? Five-digit numbers? Etc.

(v) State what you believe to be true in general. **But do not stop there!** Try to find either a *proof* which shows that your guess is correct, or a *counter example* which shows that your guess sometimes goes wrong.

5. Choose any number you like, with any number of digits—say 654321. Square each digit and add all these squares together: $6^2 + 5^2 + 4^2 + 3^2 + 2^2 + 1^2 = 91$. In this way you get a new number 91. Now do the same to your new number: $9^2 + 1^2 = 82$. Repeating this process over and over again produces a sequence of numbers:

$$654321 \rightarrow 91 \rightarrow 82 \rightarrow 68 \rightarrow 100 \rightarrow 1 \rightarrow 1 \rightarrow \ldots.$$

At this stage you may as well stop.

(i) Apply the same procedure to each of the numbers 1 to 9. Which of these starting numbers produce sequences ending in a long string of ones. What do the sequences you get from all the other numbers have in common?

(ii) Now check each of the starting numbers from 10 to 99. Does each one give rise to a sequence of one of the two types found in Part (i)?

(iii) If you could somehow show that each starting number $\geqslant 100$ produces a sequence which eventually contains a number $\leqslant 99$, then you could be sure that *every* sequence ends in one of these two ways. How could you possibly check *every* starting number $\geqslant 100$? (At first sight this seems a hopeless task. But is it really?)

6. Numbers like 7, 121, 33, 76567 which read the same both forwards and backwards are called *palindromes*.

(i) Start with the number $687_{\text{base 10}}$.
Reverse the order of the digits to get a new number. Add this to the number you started with. What answer do you get? Is it a palindrome?

Now take this answer as a new starting number, reverse the order of the digits, and add as before. What answer do you get? Is it a palindrome?

Repeat the procedure one more time. This time you should get a palindrome.

(ii) Now start with the number $79_{\text{base 10}}$ and repeat the same procedure. Do you get a palindrome eventually? How many steps does it take?

(iii) Most two-digit numbers produce a palindrome after at most two steps. Find the exceptions. How many steps do they take?

(iv) One two-digit number takes more than twenty steps to produce a palindrome. Which is it? How many steps does it take?

(v) What about numbers between 100 and 149? Do any of these require more than two steps?

(vi) What about numbers between 150 and 199? Which of these requires the most steps before producing a palindrome? (You should be able to identify this number without working out how many steps it actually takes.)

(vii) The two-digit number 89 took a surprisingly long time to produce a palindrome—much longer than any other two-digit number. And if you ignored the advice in (vi), you will have discovered that 196 is even more reluctant to produce a palindrome. At this stage you may begin to wonder whether the procedure will in fact always produce a palindrome eventually. How could one prove that it always will? Or that it sometimes doesn't? (Perhaps 196 does eventually produce a palindrome. But how could one prove that it **must**?) There are no easy answers. But once you begin to suspect that the procedure might not always work, it is worth looking around for evidence to support, or challenge, your suspicion. Where could you look?

(viii) There is nothing about our procedure which would suggest that it should work in base 10, but not in other bases. Why not test it in other bases? What is the simplest base of all? Test all one-digit, all two-digit, all three-digit, all four-digit, and all five-digit numbers in base 2. What is the first number you come up with that takes more than two steps to produce a palindrome? How many steps does it take? The next number requires one more step than this. The one after that is itself a palindrome. But how about the one after that? Look carefully at the sequence it generates.

7. Some fractions have decimals which *terminate*: for example, $\frac{1}{2} = 0.5$. All other fractions have *recurring* decimals which continue for ever with an endlessly repeating cycle of digits: for example, $\frac{1}{11} = 0.0909090\ldots$. The recurring decimals for $\frac{1}{11} = 0.0909090\ldots$, for $\frac{12}{110} = 0.10909090\ldots$, and for $\frac{10}{11} = 0.9090909\ldots$ all have, in some sense, exactly the same repeating cycle of digits—namely '09' (or, if you prefer, '90'). Similarly the recurring decimals for $\frac{1}{7} = 0.1428571428571\ldots$, for $\frac{3}{7} = 0.4285714285714\ldots$, and for $\frac{2}{7} = 0.2857142857142\ldots$ all have the same endlessly repeating cycle '142857' (or '428571', or '285714').

(i) Work out all the *different* repeating cycles which occur for fractions with denominator 11. You should find that there are exactly five different cycles, and that each of the digits 0, 1, 2, 3, 4, 5, 6, 7, 8, 9 occurs precisely once in these five cycles.

(ii) Now work out all the different repeating cycles which occur for fractions with denominator 21. Think about what you find. Then guess what you would expect to find when you work out all the different repeating cycles for fractions with denominator 31. Check your guess, and revise it as often as necessary until your predictions appear to fit the facts. Then try to prove that your guess is correct.

(iii) The only repeating cycles which occur for fractions with denominator 9 are '1', '2', '3', '4', '5', '6', '7', '8'. ($\frac{9}{9} = 0.999999\ldots = 1$, so $\frac{2}{9}$ has a decimal which *terminates*.) Thus for fractions with denominator 9, each of the digits 1, 2, 3, 4, 5, 6, 7, 8 occurs exactly once in the repeating cycles, while the digits 0 and 9 do not occur at all. Thus one cannot expect the same pattern to hold for other denominators as held for the denominators 11, 21, 31, ... But is there some similar kind of pattern? (Which denominators do you have to consider? Once you have decided which denominators to concentrate on, experiment until you think you know what happens, then try to prove it.)

Hints and references for the further problems

Hints

2. See Ref. 5, pages 73 and 78–80; Ref. 8, pages 65–73.

3. See Ref. 1, pages 31–7; Ref. 7, pages 7–8.

4. (v) Suppose your starting number N_1 has d digits, and that you apply the given procedure to get a long sequence of numbers each with d digits:

$$N_1 \to N_2 \to N_3 \to N_4 \to N_5 \to \ldots .$$

Can you somehow show that the *largest* digit in N_1 will almost always get smaller eventually? How does this help? See Ref. 5, pages 73–4 and 80–4; Ref. 6, pages 160 and 299.

5. See Ref. 3, pages 23–5; Ref. 5, pages 74 and 83–4; Ref. 6, pages 161 and 299–300.

6. (iii) Which is the first two-digit number that actually requires two steps? The next nine numbers after this are just like ones you have done already. So what is the second two-digit number you have to test? You can in fact find all the exceptions by testing precisely eight two-digit numbers. Which are they?

(v) What is the first such number that requires more than one step? And what is the next such number you really have to test?

(viii) Look especially carefully at the 5th, 9th, 13th, etc. terms in the sequence. Can you prove that this pattern continues? See Ref. 4, pages 242–44.

7. (ii) To do this you will have to sort out what it is that decides whether two occurrences of the same digit in two repeating cycles are 'indistinguishable' or 'distinct'. Why does the digit '1' in the repeating cycle of $\frac{1}{21} = 0.0476190476190\ldots$ lead on to exactly the same repeating cycle '190476' as the '1' in the repeating cycle of $\frac{4}{21} = 0.1904761904761\ldots$? What is the difference between these two occurrences of the digit '1' and the '1' in the repeating cycle of $\frac{3}{21} = 0.1428571428571\ldots$?

(iii) Why is it not worth worrying about denominators of the form $10 \times n$? Which other denominators can you safely ignore? So which denominators should you concentrate on? See Ref. 2, page 34.

References

1. A. H. Beiler *Recreations in the theory of numbers*. Dover, New York, 1966.
2. T. Beldon Recurring decimals. *Mathematics teaching* **70**, 1975.
3. J. K. Bidwell Some functions on the digits of a number. *Maths in School* **11**, 1982.
4. M. Gardner *Mathematical circus*, Pelican, Harmondsworth, Middx, 1981.
5. R. Honsberger *Ingenuity in mathematics*, Mathematical Association of America, Washington, 1970.
6. B. A. Kordemsky, *The Moscow puzzles*. Pelican, Harmondsworth, Middx, 1975.
7. W. W. Rouse-Ball and H. S. M. Coxeter, *Mathematical recreations and essays*. University of Toronto Press, Toronto, 1974.
8. P. J. van Albada and J. H. van Lint, On a Recurring Process in Arithmetic *Nieuw Archief voor Wiskunde (3)*, **IX**, 1961.

2.4 Solutions to Exercises in 2.2

Exercise 1 (i) $1078_{\text{base }9}$ (unless, of course, you started with a palindrome—in which case you would get the answer $000_{\text{base }9}$).

Exercise 2 $1056_{\text{base }7}$. Yes.

Exercise 3 (i) Yes. If $a > c$, then $abc - cba = (a - c) \times 10^2 + (c - a) = (a - c - 1) \times 10^2 + 9 \times 10 + (10 + c - a) = def$, so $d = a - c - 1$, $e = 9$, $f = 10 + c - a$. Thus $def + fed = [(a - c - 1) \times 10^2 + 9 \times 10 + (10 + c - a)] + [(10 + c - a) \times 10^2 + 9 \times 10 + (a - c - 1)] = 1089_{\text{base }10}$. (*Alternatively* casting out nines shows that abc and cba leave the same remainder on division by 9, so $abc - cba = def$ is a multiple of 9. But the middle digit e is equal to 9, so $d + f$ is a multiple of 9. Since $d = f = 9$ is clearly impossible, we must have $d + f = 9$. But then $def + fed = (d + f) \times 10^2 + 2e \times 10 + (d + f) = 1089$.)

(iii) Both methods of proof given in (i) work in any base b ($b \geqslant 2$), and show that one will always get the answer $1 \times b^3 + 0 \times b^2 + (b - 2) \times b + (b - 1)$.

Exercise 4 (i) 99. Yes. If $a > b$, then $ab - ba = (a - b) \times 10 + (b - a) = (a - b - 1) \times 10 + (10 + b - a) = cd$. So $cd + dc = 99$.

Exercise 5 There are five possible answers depending on the number $abcd$ you start with. (a) If $abcd$ is a palindrome ($a = d$ and $b = c$), you obviously get **0000**. (b) If $a = d$ and $b > c$, then $abcd - dcba = 0 \times 10^3 + (b - c - 1) \times 10^2 + (10 + c - b) \times 10 + 0 = efgh$, and $efgh + hgfe = $ **0990**. (c) If $a > d$ and $b = c$, then $abcd - dcba = (a - d - 1) \times 10^3 + 9 \times 10^2 + 9 \times 10 + (10 + d - a) = efgh$, and $efgh + hgfe = $ **10989**. (d) If $a > d$ and $b > c$, then $abcd - dcba = (a - d) \times 10^3 + (b - c - 1) \times 10^2 + (9 - (b - c)) \times 10 + (10 - (a - d)) = efgh$, and $efgh + hgfe = $ **10890**. (e) If $a > d$ and $b < c$, then $abcd - dcba = (a - d - 1) \times 10^3 + (10 + b - c) \times 10^2 + (c - b - 1) \times 10 + (10 + d - a) = efgh$, and $efgh + hgfe = $ **9999**.

3 Thinking things through

Newton, when questioned about his methods of
work, could give no other answer but that he
was wont to ponder again and again on a
subject . . . Scientists and artists both
recommend persistent labour.
(E. Mach, *Accident in invention and discovery*)

3.1 Introduction

'When is a problem *not* a problem?' The answer should be 'When it's too *easy*!' But in practice, students (and others) are more likely to reject unfamiliar or mildly confusing questions because they appear *too hard*. The problem in this third short investigation provokes precisely such a response from many students. This is a perfectly understandable reaction. It just happens to be based on a misunderstanding of what mathematics is about. Let me try briefly to explain.

Most of us learn mathematics as a collection of standard techniques which are used to solve standard problems in predictable contexts. Now techniques are important. And the mathematics student—like the athlete or musician—needs basic training to develop technique. But the athlete trains in order to *compete,* and the musician practises in order to *make music.* In the same way, the mathematician needs techniques in order to 'make mathematics' *by tackling challenging problems.*

At first sight a new piece of music may look like a confusing array of black blobs. But as the musician works on it, one section at a time, it gradually takes on a shape of its own, revealing internal connections which had previously been overlooked. The same is true of an unfamiliar mathematical problem. At first sight you may not even understand the question. But as you painfully try to make sense of the problem, you will find that, little by little, the fog begins to lift.

This experience of initial confusion, giving way as one grapples with a problem to unexpected insight, is in no way confined to beginners. It is part of the very nature of mathematics and of the way human beings *do* mathematics. The mighty Gauss (1777–1855) used 'to tell

23

his friends that if others would meditate as long and as deeply as he did on mathematical truths, they would be able to make his discoveries. He said that often he meditated for days on a piece of research without finding a solution which finally became clear to him after a sleepless night'. (G. W. Dunnington *Carl Friedrich Gauss: Titan of Science*. Hafner Publishing Co., New York, 1955).

3.2 Third short investigation

Some statements have the curious property of describing *themselves* accurately. Consider the following:

> This sentence contains thirty six letters.
> This sentence uses thirty symbols.

Making up such sentences might look like a mere game, having nothing whatever to do with mathematics. But the kind of thinking you have to use is very mathematical. And the technique of constructing special (mathematical) statements which refer to themselves has been responsible for some startling developments in mathematics during the twentieth century. The third short investigation starts out from a puzzle of this type. Try to solve the puzzle completely before reading the text that follows it.

Problem Find a list of eight whole numbers

$$n_0, n_1, n_2, n_3, n_4, n_5, n_6, n_7$$

with the property that

n_0 = the number of times 0 occurs in your list,
n_1 = the number of times 1 occurs in your list,

and so on. Is your list the only one with the required property? (If it is, explain why it is. If it is not, find another.)

The way this problem is stated tempts one to think of it as an isolated puzzle. But there are a number of reasons why it may be worth looking at other, similar, puzzles. Perhaps you found it so mind-boggling that you couldn't do it at all. Or you may have succeeded in finding one such list, but could not find a convincing proof that your list was the only possible solution. In either case it would be worth having a look at a simpler version of the same problem. But even if you managed to solve the problem completely, the mathematician in you should be mildly inquisitive as to why it asks for a list of *eight* whole numbers rather than say seven, or nine, or six hundred and thirty four. So we shall start our investigation by looking at the very simplest versions of the same problem and ask: is it possible to construct very short lists with the same property?

Exercise 1 Find a 'list' of just one whole number, n_0, with the property that n_0 is equal to the number of times 0 occurs in your list.

Exercise 2 Find a list of just two whole numbers, n_0 and n_1, with the property that n_0 is the number of times 0 occurs in the list, and n_1 is the number of times 1 occurs in the list.

Exercise 3 Find a list of three whole numbers, n_0, n_1, and n_2, with the property that n_0 is the number of times 0 occurs in the list, and so on.

You may by now be beginning to suspect that there may well have been some good reason why the original problem asked for a list of *eight* whole numbers. Still, perhaps we should keep going a bit longer before jumping to conclusions.

Exercise 4 Find all possible lists of four whole numbers, n_0, n_1, n_2, and n_3, with the property that n_0 is the number of times 0 occurs in the list, and so on. (There are at least two different lists!)

Well, where do you go from here? There is not much point marching endlessly on, increasing the length of the list by one each time in the hope of finding first all lists with five numbers, then all lists with six numbers, and so on. You might possibly be lucky and discover some pattern, completely by chance. But you are more likely to start making mistakes! For as the length of the list increases, it looks as though it is going to get more and more difficult to find all possible lists *and to be sure that you haven't missed any.*

So what should you do instead? Whatever you do, you should be on the lookout all the time for *general principles* and new *insights*. You may eventually decide that it would after all be a good idea to find all possible lists with five numbers, all possible with six numbers, and so on. But before you do, you will probably need some new ideas to simplify the calculations. The next exercise suggests that you should go back and take a fresh look at the original problem and its solution in search of such new ideas.

Exercise 5 Let n_0, n_1, n_2, n_3, n_4, n_5, n_6, n_7 be an (unknown) list of eight whole numbers with the usual property. Thus each n_i 'counts' the number of times the number 'i' occurs in the list.

(i) Could n_4, n_5, n_6, and n_7 all be zero?

(ii) How many of n_4, n_5, n_6, and n_7 can be non-zero?

(iii) Can any of the numbers n_4, n_5, n_6, and n_7 be $\geqslant 2$?

(iv) You now know that exactly one of the numbers n_4, n_5, n_6, and n_7 is equal to 1. Hence $n_1 \geqslant 1$. Could n_1 actually be equal to 1.

(v) What happens if $n_1 = 2$? Is $n_1 = 3$ possible? Can we have $n_1 \geqslant 4$?

You may have noticed that some of the ideas used in Exercise 5 look as though they should work for any list, no matter how long it happens to be.

Exercise 6 Let $n_0, n_1, n_2, \ldots, n_{99}$ be a list of one hundred whole numbers with the usual property that each n_i is equal to the number of times the number 'i' occurs in the list.

(i) Could the numbers $n_{50}, n_{51}, n_{52}, \ldots, n_{99}$ all be zero?

(ii) How many of the numbers $n_{50}, n_{51}, n_{52}, \ldots, n_{99}$ can be $\geqslant 1$?

(iii) Can any of the numbers $n_{50}, n_{51}, n_{52}, \ldots, n_{99}$ be $\geqslant 2$?

(iv) What can you say about the number n_0? What can you say about the number n_1?

(v) Can you say anything at all about the numbers $n_2, n_3, \ldots, n_{49}$?

It is beginning to look as though you should always be able to say quite a lot about the numbers in the second half of any such list. You also know quite a lot about the very first number n_0. But you have, as yet, very little idea how the other numbers in the first half of such a list will behave. These are excellent reasons why it may now be worth going back to the systematic search that was so rudely interrupted after Exercise 4. You have now discovered some general methods which should make the calculations considerably easier. And there are some specific things you would like to know. For example: which of the numbers in the second half of the list can be equal to 1? And how does the first half of each list behave?

It is often a good idea to test a new approach by using it on something you have already done in a different way. So why not start by finding all possible lists of four whole numbers all over again!

Exercise 7 Let n_0, n_1, n_2, n_3 be an (unknown) list of four whole numbers with the usual property.

(i) Could n_2 and n_3 both be zero?

(ii) Could both of the numbers n_2 and n_3 be $\geqslant 1$?

(iii) Show that if $n_3 \geqslant 1$, then $n_2 = 0$ and $n_3 = 1$. Then find n_0 and n_1.

(iv) Show that if $n_2 \geqslant 1$, then $n_2 = 1$ or $n_2 = 2$.

(v) Find n_0, n_1 when $n_2 = 1$. Find n_0, n_1 when $n_2 = 2$.

Exercise 8 Let n_0, n_1, n_2, n_3, n_4 be a list of five whole numbers with the usual property.

(i) Is there any obvious reason why n_3 and n_4 could not both be zero? Could n_2, n_3, and n_4 all be zero?

(ii) How many of the numbers n_2, n_3, and n_4 can be simultaneously $\geqslant 1$?

(iii) Could either n_3 or n_4 be $\geqslant 2$? Could n_2 be $\geqslant 2$?

(iv) Is $n_4 = 1$ possible? Is $n_3 = 1$ possible? Is $n_4 = n_3 = 0$ possible?

Exercise 9 Let n_0, n_1, n_2, n_3, n_4, n_5 be a list of six whole numbers with the usual property.

(i) Could n_3, n_4, and n_5 all be zero?

(ii) How many of the numbers n_3, n_4, and n_5 can be simultaneously $\geqslant 1$?

(iii) Can any of the numbers n_3, n_4, and n_5 be $\geqslant 2$?

(iv) Is $n_5 = 1$ possible? Is $n_4 = 1$ possible? Is $n_3 = 1$ possible?

You may by now feel that you could 'find all possible lists of *seven* whole numbers' more or less with your eyes shut. But perhaps you had better keep your eyes open, for there is as yet little sign of any general pattern. And the number of lists must start increasing again soon, because you already know that there is *precisely one such list of eight whole numbers.*

Exercise 10 Let n_0, n_1, n_2, n_3, n_4, n_5, n_6 be a list of seven whole numbers with the usual property.

(i) Could n_4, n_5, and n_6 all be zero? Could n_3, n_4, n_5, and n_6 all be zero?

(ii) How many of the numbers n_3, n_4, n_5, and n_6 can be simultaneously $\geqslant 1$?

(iii) Could any of the numbers n_4, n_5, and n_6 be $\geqslant 2$? Could n_3 be $\geqslant 2$?

(iv) Is $n_6 = 1$ possible? Is $n_5 = 1$ possible? Is $n_4 = 1$ possible? Is $n_3 = 1$ possible?

(v) Write down all possible lists n_0, n_1, n_2, n_4, n_5, n_6. How many different lists are there?

You may think you can see some connection between the list with seven whole numbers which you found in Exercise 10 and the list with eight whole numbers which you found in Exercise 5. But you would be wise to check any connection you suspect exists by finding all such lists with nine and with ten whole numbers.

Exercise 11 Let n_0, n_1, n_2, n_3, n_4, n_5, n_6, n_7, n_8 be a list of nine whole numbers with the usual property.

(i) Try to guess what you think you will find when you work out all possible such lists.

(ii) Then do the calculations carefully, using the same approach as in Exercise 10, to see if your guess was correct.

Exercise 12 (i) Use the experience of Exercises 10 and 11 to guess what you will probably find when you work out all possible lists of ten whole numbers n_0, n_1, ..., n_9 with the usual property.

(ii) Then do the calculations carefully. Was your guess correct?

Exercise 13 Whether or not you think you know what is going on, write out the lists you have found with seven, eight, nine, and ten whole numbers one under the other as in Fig. 3.1

	n_0	n_1	n_2	n_3	n_4	n_5	n_6	n_7	n_8	n_9
List with seven whole numbers										
List with eight whole numbers										
List with nine whole numbers										
List with ten whole numbers										

Fig. 3.1

You should now feel that you know what is going on. But how on earth can one be sure that this pattern really does go on for ever—with just one very special list of each length ≥ 7? You should try to think this out for yourself. When you have had a long think you might like to work through the final two exercises.

Exercise 14 Let $n_0, n_1, n_2, \ldots, n_{2m-1}$ be a list of *even* length $2m \geq 8$.

(i) Can $n_m, n_{m+1}, \ldots, n_{2m-1}$ all be zero? How many of $n_m, n_{m+1}, \ldots, n_{2m-1}$ can be simultaneously ≥ 1? Can any of $n_m, n_{m+1}, \ldots, n_{2m-1}$ be ≥ 2?

(ii) Is $n_{2m-1} = 1$ possible? Is $n_{2m-2} = 1$ possible? Is $n_{2m-3} = 1$ possible?

(iii) Is $n_{2m-4} = 1$ possible? If $n_{2m-4} = 1$, then the number '$2m - 4$' must occur in the list precisely once. Where must it occur? If $n_{2m-4} = 1$, then the number '1' occurs in the list at least once. So what must n_1 be? Does this determine the whole list?

(iv) You must now explore the only remaining possibility: namely that one of the numbers $n_m, n_{m+1}, \ldots, n_{2m-5}$ is non-zero.

Exercise 15 Let $n_0, n_1, n_2, \ldots, n_{2m}$ be a list of *odd* length $\geqslant 7$. Modify the approach outlined in Exercise 14 to show that there is exactly one such list for each value of m.

If your algebraic manipulation is quite good, you might like to look for a simple algebraic proof that there is only one possible list of each length $\geqslant 7$.

Hints to selected exercises

5. (i) Suppose they were all zero. How big would this make n_0?

(ii) If n_4 is not zero, then some number appears exactly four times in the list. What could go wrong if n_4 and n_5 were both non-zero?

14. (iii) Could the number '$2m - 4$' occur as n_1? Could it occur as n_2? What must it be? Why could we not have $n_1 = 1$?

(iv) There is only one non-zero number in the second half of the list, and this must be n_{2m-4-k} for some $k \geqslant 1$. Then $n_{2m-4-k} = 1$, so the number '$2m - 4 - k$' occurs precisely once in the first half of the list. So where must it occur? Now look at the first half of the list. How many zeros are there in the first half of the list? Let j be the largest number $\leqslant m - 1$ for which $n_j \neq 0$. Then some number has to occur j times in the list. Use the fact that there are only $j - 1$ gaps to be filled in between n_0 and n_j to show that $n_1 = j$. Hence find j.

3.3 Further problems

1. Can you construct an *endless* list of whole numbers

$$n_0, n_1, n_2, n_3, n_4, n_5, \ldots, n_k, \ldots$$

with the property that each n_i counts the number of times that the number 'i' occurs in your list?

2. The number triangle in Fig. 3.2 is meant to go on forever, with each entry being equal to the sum of the three numbers directly above it in the previous row. Will every row after the second contain at least one even number?

3. (i) Show that in a sequence of 0s and 1s with at least four terms

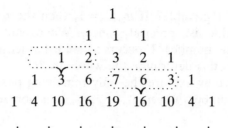

Fig. 3.2

there is always some digit, or some string of digits, which appears twice in succession.

(ii) Can you construct arbitrarily long sequences of 0s and 1s in which no digit, and no string of digits, appears three times in succession? If you think you can, show that your construction produces sequences with the required property. If you decide that it cannot be done, explain why not.

(iii) Can you construct arbitrarily long sequences of 0s, 1s, 2s, and 3s in which no digit, and no string of digits, appears twice in succession?

(iv) Can you construct arbitrarily long sequences of 0s, 1s, and 2s in which no digit, and no string of digits, appears twice in succession?

4. Start with any old list n_0, n_1, n_2, n_3, n_4, n_5, n_6, n_7. Then *improve* your list step by step as follows.

(1) Redefine n_0: = number of digits 0 in current list '–, n_1, n_2, n_3, n_4, n_5, n_6, n_7'.

(2) Redefine n_1: = number of digits 1 in current list 'n_0, –, n_2, n_3, n_4, n_5, n_6, n_7'.

(3) Redefine n_2: = number of digits 2 in current list 'n_0, n_1, –, n_3, n_4, n_5, n_6, n_7'.

(4) Redefine n_3: = number of digits 3 in current list 'n_0, n_1, n_2, –, n_4, n_5, n_6, n_7'.

(5) Redefine n_4: = number of digits 4 in current list 'n_0, n_1, n_2, n_3, –, n_5, n_6, n_7'.

(6) Redefine n_5: = number of digits 5 in current list 'n_0, n_1, n_2, n_3, n_4, –, n_6, n_7'.

(7) Redefine n_6: = number of digits 6 in current list 'n_0, n_1, n_2, n_3, n_4, n_5, –, n_7'.

(8) *Redefine* n_7: = number of digits 7 in current list 'n_0, n_1, n_2, n_3, n_4, n_5, n_6, –'.

Repeat this sequence of steps over and over again until you get a list which doesn't change. (For example, if we start with '99, 43, 76, 181, 6, 17, 29, 62', then after one run through the sequence of steps (1)–(8) you should get '0, 3, 2, 1, 0, 0, 1, 0'. After two more runs through you should get a familiar looking list, which is unaffected by further runs through the same sequence of steps.) Does this procedure always 'converge' to a list which does not change, no matter what list you start with? Does the final 'stable' list always solve the problem in Section 3.2? Can the procedure be improved?

Reference for further problem 3

3. A. M. Yaglom and I. M. Yaglom *Challenging mathematical problems with elementary solutions* Vol. 2, pp. 12 and 90–8. Holden-Day, San Francisco, 1967.

3.4 Solutions to exercises in 3.2

Exercise 1 No such list is possible. (n_0 cannot be 0 since it counts the number of times '0' occurs in the list. And n_0 cannot be greater than zero unless '0' occurs in the list.)

Exercise 2 No such list is possible. (As before n_0 cannot be zero. And if $n_0 = 1$, then '0' must occur in the list so n_1 must equal 0, which contradicts the fact that '1' has already occurred in the list.)

Exercise 3 No such list is possible. (n_0 cannot be zero. If $n_0 = 1$, then n_1 cannot be 1 so $n_1 = 2$. But then $n_2 \geqslant 1$, which contradicts the fact that '0' is supposed to occur somewhere in the list. Finally if $n_0 = 2$, then $n_2 \geqslant 1$ and there is no room for two zeros to occur in the list.)

Exercise 4 n_0 cannot be zero. Suppose first that $n_0 = 1$. Then $n_1 \geqslant 1$ and n_1 cannot be 1, so $n_1 = 2$ or $n_1 = 3$. But n_1 cannot equal 3, because a list with three 1s and at least one 3 would have no room for any 0s, contradicting $n_0 = 1$. Therefore $n_1 = 2$. So the list must contain one 0, two 1s, and at least one 2. Hence $\boldsymbol{n_0 = 1}$, $\boldsymbol{n_1 = 2}$, $\boldsymbol{n_2 = 1}$, $\boldsymbol{n_3 = 0}$. Next suppose that $n_0 = 2$. Then $n_2 \geqslant 1$. If $n_2 = 1$, then $n_1 \geqslant 1$. Now each n_i counts the number of times that the number 'i' appears in the list, and there are exactly four numbers altogether in the list, so $n_0 + n_1 + n_2 + n_3 = 4$. Hence if $n_0 = 2$, $n_2 = 1$, and $n_1 \geqslant 1$, then n_1 must in fact equal 1, which contradicts the fact that there are then two 1s in the list. If $n_2 \geqslant 2$, then $n_0 + n_1 + n_2 + n_3 = 4$, so $\boldsymbol{n_0 = 2}$, $\boldsymbol{n_1 = 0}$, $\boldsymbol{n_2 = 2}$, $\boldsymbol{n_3 = 0}$. Hence there are just two possible solutions.

Exercise 5 (i) No. (If n_4, n_5, n_6, n_7 are all zero, then $n_0 \geqslant 4$; but then at least one number $\geqslant 4$ actually occurs in the list. So n_4, n_5, n_6, n_7 cannot all be zero.)

(ii) At most one. (If n_i and n_j are both non-zero, then there must be at least one 'i' and at least one 'j' in the list—say $n_r = i$ and $n_s = j$. But each n_k counts the number of times 'k' occurs in the list, and the list contains exactly *eight numbers altogether*, so $i + j = n_r + n_s \leqslant n_0 + n_1 + n_2 + n_3 + n_4 + n_5 + n_6 + n_7 = 8$. Therefore i and j cannot both be $\geqslant 4$.)

(iii) No. (The reasoning in (ii) shows that n_5, n_6, and n_7 cannot be $\geqslant 2$, and that if $n_4 \geqslant 2$, then $n_4 = 2$. But if $n_4 = 2$, then there are two 4s *and at least one* 2, so the sum of all the n_is would be at least 10.)

(iv) No.

(v) Precisely one of n_4, n_5, n_6, n_7 is equal to 1, the rest are 0. So $n_0 \geqslant 3$. By (iv) we know that $n_1 \geqslant 2$. If $n_1 = 2$, then this tells us both that $n_2 \geqslant 1$ and that the list must contain exactly two 1s, so either n_2 or n_3 must be equal to 1. If $n_2 > 1$, then $n_3 = 1$ and the sum of all the n_is would be at least 9. Therefore $\boldsymbol{n_2 = 1}$, and $n_3 \neq 1$ so $\boldsymbol{n_3 = 0}$, $\boldsymbol{n_0 = 4}$, $\boldsymbol{n_4 = 1}$, $\boldsymbol{n_5 = n_6 = n_7 = 0}$. If $n_1 \geqslant 3$, then we need at least two more 1s, so the only possibility would be $n_1 = 3$, $n_2 = n_3 = 1$. But then $n_0 = 3$ and the list contains two 3s, contradicting $n_3 = 1$.

Exercise 6 (i) No. (If $n_{50} = n_{51} = n_{52} = \ldots = n_{99} = 0$, then $n_0 \geqslant 50$ so at least one number $\geqslant 50$ occurs in the list. But then $n_{50}, n_{51}, \ldots, n_{99}$ cannot all be zero.)

(ii) At most one (for the same reason as in Exercise 5(ii)).

(iii) None (for the same reason as in Exercise 5(iii)).

(iv) Exactly forty nine of the numbers $n_{50}, n_{51}, \ldots, n_{99}$ are zero, so $n_0 \geqslant 49$. Exactly one of these numbers is equal to 1, so $n_1 \geqslant 1$. And $n_1 = 1$ is impossible (since then at least two 1s occur in the list), so $n_1 \geqslant 2$.

(v) It looks as though most of them will have to be zero, so n_0 will be quite large. But it is not easy to be much more precise than this.

Exercise 7 (i) No.

(ii) No.

(iii) If $n_3 \geqslant 1$, then $n_2 = 0$ by (ii). And $n_3 \neq 2$ for the same reason that $n_4 \neq 2$ in Exercise 5(iii); so $n_3 = 1$. Now $n_1 \geqslant 2$ (as in Exercise 6(iv)), and $n_0 \geqslant 1$, so $n_0 = 1$, $n_1 = 2$, $n_2 = 0$, $n_3 = 1$.

(iv) $n_2 \neq 3$ for the same reason that $n_4 \neq 2$ in Exercise 6(iv), so $n_2 = 1$, or $n_2 = 2$.

(v) $n_0 \geqslant 1$. If $n_2 = 1$, then $n_1 \geqslant 2$ (as in Exercise 6(iv)), and $n_3 = 0$ by (ii). So $n_0 = 1$, $n_1 = 2$, $n_2 = 1$, $n_3 = 0$. If $n_2 = 2$, then $n_3 = 0$ by (ii). But $n_0 + n_1 + n_2 + n_3 = 4$, so the list contains exactly two 2s and two 0s. Hence $n_0 = 2$, $n_1 = 0$, $n_2 = 2$, $n_3 = 0$.

Exercise 8 (i) At first sight, yes. No.

(ii) Precisely one.

(iii) No. Suppose $n_2 \geqslant 2$. Then $n_2 = 2$ (otherwise the list would contain at least three 2s, so the sum of all the n_is would be too big). If $n_2 = 2$, then $n_3 = n_4 = 0$ by (ii), so $n_0 \geqslant 2$ and we must have $n_0 = 2$, $n_1 = 1$, $n_2 = 2$, $n_3 = 0$, $n_4 = 0$.

(iv) If $n_4 = 1$, then some number must occur four times in the list. But $n_1 \geqslant 1$ and $n_2 = n_3 = 0$ by (ii), so $n_0 \geqslant 2$ and no number can occur four times in the list. If $n_3 = 1$, then some number must occur three times in the list. But $n_1 \geqslant 1$ and $n_2 = n_4 = 0$ by (ii), so $n_0 \geqslant 2$ and no number occurs three times in the list. If $n_3 = n_4 = 0$, then $n_2 \geqslant 1$ by (ii). If $n_2 \geqslant 2$, then we get the list in (iii). Suppose $n_2 = 1$. Then $n_3 = n_4 = 0$ so $n_0 \geqslant 2$, and since the list contains no 3s and no 4s we have $n_0 = 2$. But then n_1 goes wrong. ($n_1 \geqslant 1$ since the list contains at least one 1, and $n_1 \neq 1$, so since the list contains no 3s or 4s we must have $n_1 = 2$. But then the list would contain two 2s, contradicting $n_2 = 1$.)

Exercise 9 (i) No.

(ii) Precisely one.

(iii) No.

(iv) (a) If $n_5 = 1$, then some number must occur exactly five times in the list. But $n_1 \geq 1$, and $n_3 = n_4 = 0$ by (ii), so $n_0 \geq 2$ and no number can possibly occur five times in the list. (b) If $n_4 = 1$, then some number must occur exactly four times in the list. But $n_1 \geq 1$, and $n_3 = n_5 = 0$ by (ii), so $n_0 \geq 2$ and no number can possibly occur four times in the list. (c) If $n_3 = 1$, then some number must occur exactly three times in the list. Also $n_4 = n_5 = 0$ by (ii), so $n_0 \geq 2$. And $n_1 \geq 1$, but $n_1 \neq 1$, so $n_1 \geq 2$. But the only way to get at least two 1s in the list is to put $n_2 = 1$, and this would imply that $n_0 = 2$, $n_1 = 2$ so the list would contain no 3s at all.

Exercise 10 (i) Yes. No.

(ii) Precisely one.

(iii) No. No. (Suppose $n_3 \geq 2$. Then either $n_3 \geq 3$ and the list contains at least three 3s, or $n_3 = 2$ and the list contains at least two 3s and one 2. In each case the sum of all the n_is would be too big.)

(iv) No (for the same reason as in Exercise 9(iv)(a)). No (for the same reason as in Exercise 9(iv)(b)). No (for the same reason as in Exercise 9(iv)(c)). Finally suppose $n_3 = 1$. Then $n_4 = n_5 = n_6 = 0$ so $n_0 \geq 3$, and since the list contains no 4s, 5s, or 6s we must have $n_0 = 3$. Since $n_3 = 1$ we must have $n_1 \geq 1$, and $n_1 \neq 1$, so $n_1 \geq 2$. But only one 3 is allowed, so $n_1 = 2$. Then $n_2 = 1$ and we get the list $n_0 = 3$, $n_1 = 2$, $n_2 = 1$, $n_3 = 1$, $n_4 = n_5 = n_6 = 0$.

(v) Just one (see (iv)).

Exercise 11 (ii) Precisely one of n_4, n_5, n_6, n_7, n_8 is equal to 1 and all the rest are zero. (Why is $n_4 = 2$ impossible?) Thus $n_0 \geq 4$ and $n_1 \geq 2$, so $n_1 = 2$ or $n_1 = 3$. $n_6 \neq 1$ since there is not enough room left over for any number to occur six times in the list. Similarly $n_7 \neq 1$ and $n_8 \neq 1$. If $n_5 = 1$, then $n_4 = n_6 = n_7 = n_8 = 0$. If $n_1 = 2$, then $n_2 = 1$, $n_3 = 0$, $n_0 = 5$. If $n_1 = 3$, then $n_2 = 0$, $n_0 = 5$ and the list cannot possibly contain three 1s as required. Finally suppose that $n_4 = 1$. Then $n_5 = n_6 = n_7 = n_8 = 0$, so $n_0 \neq 5$. Therefore $n_0 = 4$, so neither n_2 nor n_3 can be zero. But then the list includes one 4, at least two 1s, at least one 2, and at least one 3, so the sum or all the n_is is too big.

Exercise 13

	n_0	n_1	n_2	n_3	n_4	n_5	n_6	n_7	n_8	n_9
List with seven whole numbers	3	2	1	1	0	0	0			
List with eight whole numbers	4	2	1	0	1	0	0	0		
List with nine whole numbers	5	2	1	0	0	1	0	0	0	
List with ten whole numbers	6	2	1	0	0	0	1	0	0	0

Exercise 14 (i) No. Precisely one. No.

(ii) No. No. No.

(iii) Yes. If $n_{2m-4} = 1$, then the sum of all the n_is includes the number $2m - 4$ once and some other number $2m - 4$ times. Since this sum must equal $2m$, the number which occurs $2m - 4$ times must be 0. Hence $n_0 = 2m - 4$. Also $n_1 \geqslant 1$ and $n_1 \neq 1$, so $n_1 \geqslant 2$. But then the sum of all the n_is includes $2m - 4$ once, 1 at least twice, and some number 2 at least once. Since this sum must equal $2m$ we have to have $n_1 = 2$, $n_2 = 1$, $n_3 = n_4 = \ldots = n_{2m-5} = 0 = n_{2m-3} = n_{2m-2} = n_{2m-1}$.

(iv) Suppose $n_{2m-4-k} = 1$ for some $k \leqslant m - 4$. Then the number $2m - 4 - k$ occurs in the first half of the list, so one of the numbers $0, 1, \ldots, m - 1$, say j, occurs $2m - 4 - k$ times. The sum of all the n_is thus includes $2m - 4 - k$ once and some other number j occurring $2m - 4 - k$ times. So $j = 0$ and $n_0 = 2m - 4 - k$ (otherwise the sum of all the n_is would be too big). Let p be the largest number $\leqslant m - 1$ for which $n_p \neq 0$. Now $n_1 \geqslant 2$ (since $n_{2m-4-k} = 1$ so $n_1 \geqslant 1$, and $n_1 = 1$ is self contradictory). So p must be $\geqslant 2$. Since $n_p \neq 0$, some number i must occur precisely p times in the list. If $i > 1$, then all of these p different occurrences of the number i have to be fitted in between n_1 and n_{p-1}, which is obviously impossible. Therefore $i = 1$ and $n_1 = p$, so the number 1 must occur p times in the list. But there is only one non-zero number beyond n_p, so all $p - 1$ of the numbers $n_2, n_3, \ldots, n_p$ must equal 1. Hence $p = 2$. Finally since the sum of all the n_is must equal $2m$ we get $k = 0$.

4 Asking questions

When we know a fact about polygons,
we should try to discover an analogous
fact about polyhedra; in so doing we
have a good chance to hit upon a
stimulating question.
(G. Polya, *Mathematical discovery* Vol. 2)

4.1 Introduction

Most students think of Mathematics in terms of answering *other people's* questions. This is a feature of the educational process of 'being taught' rather than of Mathematics itself. In this chapter I shall try to show you a different aspect of Mathematics—namely what can happen when *you yourself* start asking the questions. I shall begin with two examples from secondary school mathematics which may well have occurred to you before.

1. (*a*) The graph of $y = x^2$ is symmetrical about the y-axis. The graphs of other quadratics like $y = 2x - 4 + x^2$ always seem to be symmetrical about the axis through the lowest (or highest) point on the curve. Is this always true? Why?
(*b*) The graph of $y = x^3$ is symmetrical under a half turn about the origin. You may have noticed that other cubic curves, such as $y = x^3 - x$ or $y = 7x^2 - x + x^3 + 2$, often seem to be symmetrical under a half turn about a point half way between the maximum and the minimum. Is this always true? Why?
(*c*) What about curves of higher degree?

2. (*a*) What is the connection between the formulae $2\pi r$ for the *circumference* and πr^2 for the *area* of a circle of radius r? Is there a similar connection between the formulae for the surface area and the volume of a sphere? Why?
(*b*) Why does the same constant π appear both in the formula for the area of a (two-dimensional) circle and in the formula for the volume of a (three-dimensional) sphere? How big is the 'inside' of a 'four-dimensional sphere' of radius r? Does the corresponding formula have a 'π' in it? (What about a 'one-dimensional sphere of radius r'?)

Some questions are reasonably clear—even if we can't see how to answer them. (Are all cubics symmetrical under a half-turn? Are all quartics symmetrical about a vertical axis?) Other questions start out more in the spirit of a speculative 'What if . . . ?', probing for extensions of, or connections between, familiar bits of mathematics. These are bound to be a bit vague or confusing at first. (What on earth is a 'four-dimensional sphere of radius r'? And how does one measure its 'inside'?) They may nevertheless be good questions to ask.

But no matter whether the question one is trying to answer is clear or confusing, one way of getting started is to examine more carefully how one resolves related, but easier, problems. (Why are quadratics always symmetrical about a vertical axis? Can one use the same idea for cubics? And quartics? What do the formulae for the 'insides' of one, two, and three-dimensional 'spheres of radius r' suggest about the formula for the 'inside' of a 'four-dimensional sphere of radius r'? Is there a way of deriving the formula for the volume of a three-dimensional sphere which gives us a clue about how to proceed in four-dimensions?) In particular, every 'bright idea' should be *tested* on carefully chosen examples to see whether it holds good. (Thus, for example, $y = x^4$ and $y = x^4 - x^2$ may be beautifully symmetrical about a vertical axis, but how about $y = x^4 - x^3$?)

4.2 Fourth short investigation

A parallelogram can be constructed by

(i) first drawing two parallel line segments of equal length (Fig. 4.1(i)),

(ii) then joining up corresponding ends (Fig. 4.1(ii)).

If the parallel line segments we start with have length a, then all cross sections parallel to these two line segments have length a. And if the perpendicular distance between the two parallel line segments we start with is h, then the area $\mathcal{A}$ of the parallelogram is given by $\mathcal{A} = a \times h$.

(i) (ii)

Fig. 4.1

(i)   (ii)

Fig. 4.2

We can construct a trapezium in a similar way by

(i) first drawing two parallel line segments (whose lengths need not be equal this time) (Fig. 4.2(i)),

(ii) then joining up corresponding ends (Fig. 4.2(ii)).

Exercise 1 (i) Suppose that the parallel line segments have lengths a and b, and that the perpendicular distance between them is h (Fig. 4.2(ii)). Find a formula expressing the area $\mathcal{A}$ of the trapezium in terms of a, b, and h.

(ii) What shape do you get if a is zero but b is not? Does your formula for the area $\mathcal{A}$ work in this case? What if both a and b are zero? Does your formula still work?

We can now state the question to be investigated in this chapter.

Problem Is there a natural three-dimensional version of the formula for the area of a trapezium?

At first sight it is not at all clear what this is supposed to mean. The natural three-dimensional version of 'area' is presumably 'volume'. *But it is not so easy to see what one should choose as the natural three-dimensional version of a 'trapezium'.*

One would hardly expect a sphere to count as a 'three-dimensional version of a trapezium'. But how about a frustum of a square pyramid (Fig. 4.3(i))? This surely has the same sort of feel about it as a

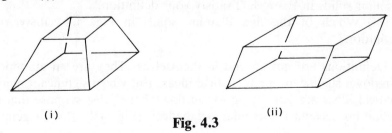

(i)   (ii)

Fig. 4.3

trapezium. And what about a parallelepiped (Fig. 4.3(ii))? After all, a parallelogram is a special kind of trapezium, and a parallelepiped is a 'three-dimensional version of a parallelogram'.

Exercise 2 Try to draw three other familiar solids which you feel deserve to be called 'three-dimensional versions of a trapezium'. What is it about a frustum of a square pyramid, a parallelepiped, and your other three familiar solids which makes it reasonable to think of each one as a 'three-dimensional version of a trapezium'?

There is no cut and dried way of deciding what does and what does not deserve to be called a 'three-dimensional version of a trapezium'. In our context one must bear in mind the way one constructs a trapezium by first drawing two parallel edges. It thus seems reasonable to suggest that a 'three-dimensional version of a trapezium' should be a solid constructed by first drawing two parallel faces and then joining them up in some way. But what shapes should these 'faces' be? And how exactly should they be 'joined up'?

To get started you will simply have to make some kind of first *guess* as to what does and what does not deserve to be called a 'three-dimensional version of a trapezium'. You can then stand back and *test* whether this first guess has the properties you would like. In particular you should check whether it allows you to prove a three-dimensional version of the formula $\mathscr{A} = \frac{1}{2}(a + b) \times h$ for the area of a trapezium. Your first guess may turn out to be *too restricted*: it may exclude certain shapes which have, on reflection, every right to be included. Or it may be *too general*: it may include certain undesirable aliens. In either case you will then want to modify, or *improve* your original guess. This process of testing and improving your first guess has to be repeated as often as necessary until you get a version which looks and feels 'right'.

Exercise 3 (i) Try to write down a definition of what you think deserves to be called 'a three-dimensional version of a trapezium'.

(ii) Now subject your definition to a first simple test. Do the five familiar solids in Exercise 2 satisfy your definition?

(iii) Which of the five Platonic solids in Fig. 4.4 satisfy your definition?

Definitions are not an end in themselves. They are an attempt to pin down significant mathematical ideas. But which significant mathematical ideas are you trying to pin down here? Just suppose that **my** definition accepts the regular octahedron (Fig. 4.4(c)) as a genuine

(a)　　　　　　(b)　　　　　　(c)　　　　　　(d)　　　　　　(e)

Fig. 4.4

'three-dimensional version of a trapezium' whereas **yours** does not. (Does your definition really reject the regular octahedron? Are you sure?) How could we choose between the two rival definitions? There is no hard and fast rule. The proof of the pudding is in the eating. In the context of this chapter the essential features of a trapezium flow from

(1) the fact that it has two parallel edges, and

(2) the very simple formula for its area.

And it is these same two features which make it so convenient to measure the areas of more complicated shapes by cutting them up into trapezia (e.g. when one calculates the area of awkward-shaped fields or other pieces of land, or when one approximates the area under a graph using the trapezium rule). So it is the three-dimensional analogues of these two features which your definition should be trying to pin down.

Before jumping to unnecessarily crude conclusions about what does and what does not deserve to be called a 'three-dimensional version of a trapezium', it may be a good idea to stand back and think about a *special case* first. Rectangles and parallelograms are two special kinds of trapezium in which the two parallel edges one starts with have the same length a. As a result the formula for the area of a rectangle or parallelogram

$$\mathscr{A} = \tfrac{1}{2}(a + b) \times h = a \times h$$

is even simpler than the general formula for the area of a trapezium. The next three Exercises and accompanying text consider the question of which solids deserve to be called 'three-dimensional versions of *rectangles*'. This should help you to take a fresh look at your original answer to Exercise 3.

Exercise 4 Think of a parallelogram as being constructed by first drawing two parallel edges of equal length, and then joining up corresponding ends. When will this procedure actually produce a *rectangle*?

Exercise 5 (i) You have probably got used to thinking of the formula for the area of a rectangle as

$$\text{area} = \text{length} \times \text{breadth.}$$

What is the natural three-dimensional version of this formula? What would the natural interpretation of 'three-dimensional version of a rectangle' be *from this point of view*?

(ii) But in the context of this present chapter a rectangle is a special kind of trapezium, constructed by first drawing two parallel *edges* of equal *size,* one directly above the other. So *from the point of view of this chapter,* the right way to think about the formula for the area of a rectangle is

$$\text{area} = (\textit{size of the equal and parallel edges})$$

$$\times (\text{perpendicular distance between them}). \qquad (4.1)$$

What formula does this suggest for the volume of a 'three-dimensional version of a rectangle'? What would the appropriate interpretation of the expression 'three-dimensional version of a rectangle' be *from this point of view*?

Even if one guessed the right kind of formula for the volume of a 'three-dimensional version of a rectangle' in Exercise 5(ii), the answer to the final question is not entirely obvious. If we think of a (two-dimensional) rectangle as being constructed by first drawing two parallel *edges* of equal *size,* one directly above the other, and then joining them up, it would seem reasonable to construct a 'three-dimensional version of a rectangle' by first drawing two parallel *faces* of equal size, one directly above the other, and then joining them up. The point of view adopted in the first part of Exercise 5 effectively forces these parallel faces to be *rectangles.* But the point of view adopted in the second part of Exercise 5 (which is the point of view of this chapter) does not seem to impose any restriction at all on the shapes of these faces, except that they should be 'of equal size' with 'one directly over the other'. The first of these restrictions ('of equal size') suggests that the two faces should both have the same *area A*; the second restriction ('one directly over the other') then seems to indicate that the two parallel faces should in fact be *congruent* to one another, and that the two faces should also be *similarly situated* (that is, the top face should not be rotated relative to the bottom face).

The idea that the two parallel faces must not only be congruent, but must also be similarly situated, should perhaps be tested in some way. What could possibly go wrong if the top face were twisted relative to the bottom face? Remember that we want to define a 'three-dimensional version of a rectangle' to be a solid whose volume $\mathcal{V}$ is given by the three-dimensional version of the formula (4.1): that is, we want the volume to be given by $\mathcal{V} = A \times h$.

Exercise 6 A solid is constructed by drawing two parallel 2 by 1 rectangles (so $A = 2$), one directly over the other, but with the top face rotated 90° relative to the bottom face (Fig. 4.5). Find its volume $\mathcal{V}$. Is it equal to, greater than, or less than $A \times h$?

So in the context of this present chapter it looks as though the expression 'three-dimensional version of a rectangle' should be taken to mean

a solid with two parallel congruent faces which are 'similarly situated', one directly over the other, these two faces being joined up in the obvious way.

Such a solid is called a **right prism.** There is no restriction at all on the shape of the two parallel faces—as long as these faces are congruent and similarly situated. It is not hard to convince yourself that the volume $\mathcal{V}$ will always be given by the formula $\mathcal{V} = A \times h$. This is the aim of the next exercise.

Exercise 7 Suppose you already know that the volume $\mathcal{V}$ of a cuboid or rectangular box is given by the formula $\mathcal{V} = \text{length} \times \text{breadth} \times \text{height}$.
(i) Use this to show that when the top and bottom faces of a right prism are rectangles of area A, the volume $\mathcal{V}$ of the prism is given by the formula $\mathcal{V} = A \times h$.
(ii) Use (i) to show that when the top and bottom faces of a right prism are parallelograms of area A, the volume of the prism is given by the formula $\mathcal{V} = A \times h$.
(iii) Use (ii) to show that when the top and bottom faces of a right prism are triangles of area A, the volume of the prism is given by the formula $\mathcal{V} = A \times h$.
(iv) Use (iii) to show that when the top and bottom faces of a right

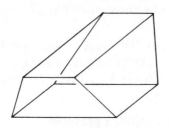

Fig. 4.5

prism are polygons with n sides having area A, the volume of the prism is given by the formula $\mathcal{V} = A \times h$.

(v) How can one use Part (iv) to show that the volume of a right circular cylinder is still given by the formula $\mathcal{V} = A \times h$?

If we now return to the original problem of trying to find a three-dimensional version of the formula for the area of a trapezium, then it is natural to expect that every right prism (that is, every three-dimensional version of a *rectangle*) should be a special case of a 'three-dimensional version of a *trapezium*'.

Exercise 8 Modify the definition of a 'right prism' which appears just before Exercise 7 to give what you would expect to be the definition of a 'three-dimensional version of a trapezium'.

There may well still be one or two awkward features of the definition you gave in your solution to Exercise 8. For example, the definition of a right prism imposed no restriction at all on the shape of the two parallel faces. This was alright as long as the two faces had to be 'congruent, with one directly above the other'. In particular, there is an 'obvious' way of joining the top face of a right prism to the bottom face (by drawing vertical line segments between corresponding points round the edges of the two faces). But what if the top face is an ellipse and the bottom face is a triangle? Which points should be joined to which? If you think about this for a while you will begin to see that these are by no means silly questions.

Finding a three-dimensional version of the formula for the area of a trapezium is bound to be harder than finding the three-dimensional version of the formula for the area of a rectangle. And whenever you go from an easy problem to a harder one you should always consider whether it might not be a good idea to impose some kind of simplifying restriction—at least in the first instance. Is there some natural restriction on the shapes of the two parallel faces which will somehow get round this problem of how the top and bottom faces should be joined up? Fortunately there is. So to get started we shall interpret the expression 'three-dimensional version of a trapezium' to mean

a solid with two parallel faces, which are both *polygons*, and which are joined corner to corner by straight edges in such a way that the resulting solid is a *polyhedron* (that is, all its faces are flat).

Any such solid will be called a **prismoid**.

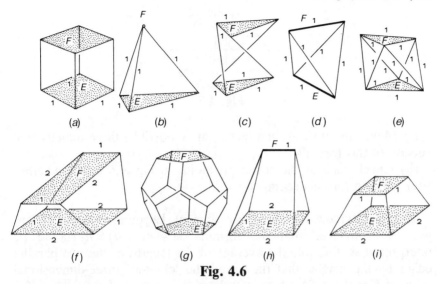

Fig. 4.6

Way back in Exercise 1 we stressed the fact that if one or both of the parallel edges in a trapezium are shrunk to a point, then the area formula still works: so it is reasonable to think of a triangle as a special kind of trapezium. In the same spirit you should be prepared to accept prismoids in which one or both of the two parallel faces have been shrunk to an edge, or even to a point.

Exercise 9 Each of the solids illustrated in Fig. 4.6 has two parallel faces E and F (shaded). All but two of these solids are prismoids. Which are the two odd men out?

Exercise 10 (i) Use any method you can think of to calculate the volume of the seven prismoids in Fig. 4.6. (For example, each one can be cut up into pieces so that each piece is either (1) a cuboid or half a cuboid, or (2) a tetrahedron or a pyramid. So you only need to know how to calculate the volume of a cuboid and the volume of a tetrahedron or pyramid.)

(ii) In each of these seven prismoids find the area A, B of the two parallel faces E, F, and the perpendicular distance h between them. (*The letters A, B, and h will have the same meaning for the rest of this chapter.*)

Exercise 11 The area of a trapezium is given by the formula $\mathscr{A} = \frac{1}{2}(a + b) \times h$, where a, b are the lengths of the two parallel edges and h is the perpendicular distance between them.

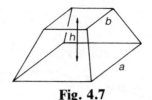

Fig. 4.7

(i) Make an obvious first guess at a possible three-dimensional version of this formula.

(ii) Check each of the seven prismoids in Fig. 4.6 to see whether your guessed formula seems to work.

There is evidence that certain ancient Egyptian mathematicians were so convinced that the formula $\mathscr{A} = \frac{1}{2}(a+b) \times h$ should be interpreted as 'multiply the average of the lengths of the two parallel edges by the height' that they used the 'obvious' three-dimensional version $\mathscr{V} = \frac{1}{2}(A+B) \times h$ to calculate the volume of a frustum of a square pyramid.

Exercise 12 A frustum of a square pyramid (Fig. 4.7) has height h. The two parallel square faces have sides of length a and b respectively. Find a formula for its volume. Does your formula look anything like $\frac{1}{2}(a^2 + b^2) \times h$?

The formula for the volume of a tetrahedron or pyramid may have suggested as a wild second guess that perhaps the three-dimensional version of $\mathscr{A} = \frac{1}{2}(a+b) \times h$ is $\mathscr{V} = \frac{1}{3}(A+B) \times h$. This will certainly work for the tetrahedron or pyramid (for then $B = 0$ and $\mathscr{V} = \frac{1}{3}A \times h$), but it will not work for the cube or any other right prism (where $A = B$ and $\mathscr{V} = A \times h$), or for the prismoid in Exercise 12.

Exercise 13 Explain why a formula which refers only to the areas A, B of the two parallel faces E, F cannot possibly work for all prismoids.

Exercise 14 The area formula $\mathscr{A} = \frac{1}{2}(a+b) \times h$ for a trapezium (Fig. 4.8) can be interpreted in a different way, namely

area = (length of cross-section midway between
the two parallel edges) × height

(i) State the three-dimensional version of this interpretation of the formula for the area of a trapezium.

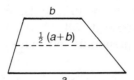

Fig. 4.8

(ii) Does this formula work for the frustum of a square pyramid in Exercise 12? Does it work for any of the prismoids in Fig. 4.6?

If a and b are the lengths of the two parallel edges of a trapezium, and $c = \frac{1}{2}(a + b)$ is the length of the cross section midway between these two parallel edges, then Exercises 11 and 14 are based on the observations that $\mathcal{A} = \frac{1}{2}(a + b) \times h = c \times h$. These exercises also show that neither of these formulae can be generalized to give a formula for the volume of a prismoid. But the fact that $c = \frac{1}{2}(a + b)$ means that there are not just two, but infinitely many different ways of writing the formula $\mathcal{A} = \frac{1}{2}(a + b) \times h$ *in terms of a, b, c, and h.* Before abandoning this line of thought it may be worth looking at one or two of these other expressions.

Exercise 15 (i) Here are the two simplest ways of writing the formula for the area $\mathcal{A}$ of a trapezium using a, b, h, *and* c. Try to find the next three simplest expressions.

$$\mathcal{A} = \tfrac{1}{2}(a + b) \times h = \tfrac{1}{3}(a + c + b) \times h = \ldots .$$

(ii) Now look at my solution to Part (i), just to make absolutely certain that we are both talking about the same expressions! The three-dimensional version of each of these formulae expresses the volume $\mathcal{V}$ in terms of the height h and the areas B, A, and C of the top face, the bottom face, and the cross-section midway between the top and bottom faces. Check that the three-dimensional version of each of your formulae works for any right prism.

(iii) Explain why at most one of these formulae could work for every prismoid.

(iv) Show that precisely one of these formulae works for the frustum of a square pyramid in Exercise 12. Does the same formula work for the other prismoids in Fig. 4.6? Does it work for right prisms? Does it work for tetrahedra and pyramids?

You should now be in the position of having a formula which seems to work for a large variety of prismoids. But will it work for all prismoids? And just suppose it will. How on earth could you *prove* that it will work for all prismoids, no matter how complicated they may be?

Exercise 16   (i) Show that a prismoid with ≤3 vertices always has $A = B = C = \mathcal{V} = 0$. Hence show that your formula works for any such prismoid.

(ii) Show that there are exactly two different 'types' of prismoid with 4 vertices. One type (Fig. 4.6(b)) has $B = 0$, $A \neq 0 \neq C$, $\mathcal{V} \neq 0$ (or $A = 0$, $B \neq 0 \neq C$, $\mathcal{V} \neq 0$); the other type (Fig. 4.6(d)) has $A = B = 0$, $C \neq 0$, $\mathcal{V} \neq 0$. Prove that your formula works for all prismoids with four vertices.

(iii) Use (ii) to prove that your formula works for any prismoid, no matter how complicated it may be.

Hints to selected exercises

6. Try dissecting the solid into simpler pieces (Fig. 4.9).
7. (ii) Chop the prism into two pieces and put them back together again to make a right prism whose top and bottom faces are rectangles with area A.

(iii) Fit two identical copies of the given triangular prism together to make a right prism whose top and bottom faces are parallelograms with area $2 \times A$.

(iv) Chop the given prism into triangular prisms. Then use (iii).
9. One of the shapes is not even a polyhedron.
10. (i) (a), (b) No problem. (d) You've just done this one. (e) Two square-based pyramids stuck base to base. (f) Exercise 6. (h) Two pyramids and half a cuboid. (i) The difference of two pyramids.
13. A prismoid with $A = B = 0$ may have $\mathcal{V} = 0$ (for example, if E and F both shrink to a point). But it can also have $\mathcal{V} \neq 0$ (as in Fig. 4.6(d)).
15. (iii) If $\mathcal{V} = \frac{1}{m}(A + (m - 2)C + B)h = \frac{1}{n}(A + (n - 2)C + B)h$ for every prismoid, then $(m - n)(A + B) = (m - n)2C$ for every prismoid.
16. (ii) Use the formula for the volume of a tetrahedron.

(iii) Show that if a prismoid with ≥4 vertices has top and bottom faces F and

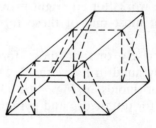

Fig. 4.9

E, then it can be cut up into prismoids each with exactly 4 vertices, with their top faces fitting together to make *F*, their bottom faces fitting together to make *E*, and their cross-sections midway between top and bottom having total area *C*. Then use (ii).

4.3 Further problems

1. (i) Can one draw an equilateral triangle with all three vertices on the dots of the square dot lattice (Fig. 4.10(i))?
(ii) Which regular polygons can one draw with all their vertices on the dots of the square dot lattice?
2. (i) Can one draw a square with all four vertices on the dots of the (equilateral) triangular dot lattice (Fig. 4.10(ii))?
(ii) Which regular polygons can one draw with all their vertices on the dots of the equilateral triangular dot lattice?
3. (i) Given a triangle *ABC*, mark points *X*, *Y*, *Z* one third of the way along the edges *BC*, *CA*, *AB*, and then join *AX*, *BY*, *CZ* (Fig. 4.11). What fraction of the area of the whole triangle *ABC* is the shaded central triangle? (Look at a special case first. You can then use simple methods to get some idea of what *answer* to expect, even if the *methods* you use do not work for a general triangle.)
(ii) Given a triangle *ABC*, join each vertex to a point $(1/n)^{th}$ of the way along the opposite edge. These three lines enclose a central triangle. What fraction of the area of the original triangle *ABC* is this central triangle? What happens when $n = 2$? What happens if $n < 0$? What happens as $n \to \infty$?
(iii) Can you devise a similar kind of result for quadrilaterals? (It is natural to begin by looking at certain special kinds of quadrilaterals first. But at some stage you will have to decide whether there is any

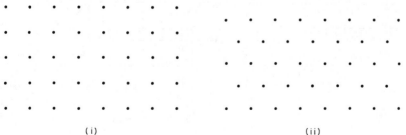

(i)   (ii)

Fig. 4.10

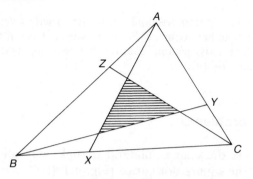

Fig. 4.11

hope of finding one single result which works for all quadrilaterals. If not, you will have to decide what the correct generalization of 'triangle' is here.)

4. (i) What is the longest straight pole that can be manoeuvred round the right-angle bend in Fig. 4.12(i)?

(ii) (a) What is the area of the largest 2-dimensional shape you can manoeuvre round a right-angle bend like the one in Fig. 4.12(i)? (This is obviously a much harder question than the one in Part (i). What extra assumption would make it easier to get started? When you have decided that, try some familiar shapes. What is the biggest rectangle you could get round? What other familiar shapes could you try?)

(b) How good is your answer to Part (a)? Can you find some obvious 'upper bound' which restricts the size of a shape which can be manoeuvred round such a corner? (In other words, can you find a number X such that any shape which can be successfully manoeuvred round the right-angle bend would have to have area $\leqslant X$?)

(iii) (a) What is the longest straight pole that can be manoeuvred round the three-dimensional right-angle bend in Fig. 4.12(ii)?

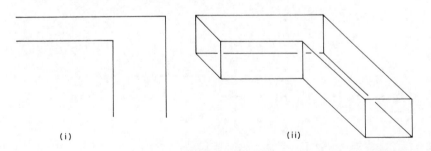

(i)   (ii)

Fig. 4.12

  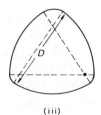

(i)   (ii)   (iii)

Fig. 4.13

(*b*) What is the volume of the largest shape you can manoeuvre round a right-angle bend like the one in in Fig. 4.12(ii)?

5. The most obvious example of a *curve of constant breadth D* is the circle of diameter *D*.

(i) The Reuleaux triangle is constructed from three 60° arcs of circles of radius *D* (Fig. 4.13(i)). Convince yourself that it is indeed a curve of constant breadth *D*. What is its perimeter? What is its area?

(ii) The British 50p coin is constructed from seven equal circular arcs of radius *D* (Fig. 4.13(ii)). It is a curve of constant breadth *D*. What is its perimeter? What is its area?

(iii) The curve in Fig. 4.13(iii) is constructed from six circular arcs—the sum of the radii of opposite arcs being always equal to *D*. It is a curve of constant breadth *D*. What is its perimeter? What is its area?

(iv) Imagine all possible curves of constant breadth *D*. What do you expect to be true about the minimum and maximum perimeter which such a curve could have? What can you say about the minimum and maximum area such a curve could have?

(v) Let γ be any curve of constant breadth *D*.

(a) Show that each tangent to γ touches the curve at just one point.

(*b*) If *a* and *b* are parallel tangents touching the curve γ at *A* and *B*, show that *AB* is perpendicular to both *a* and *b*. What does this tell you about the length of *AB*?

(*c*) As the diameter *AB* rotates through an angle $\delta\theta$, its ends move perpendicular to *AB* and trace out two infinitesimal circular arcs with the same instantaneous centre. What is the combined length of these two circular arcs? What does this tell you about the perimeter of γ?

6. How big is the 'inside' of a four-dimensional sphere of radius *r*? What about a five-dimensional sphere?

7. A circular coin with a small hole at its centre stands vertically on a horizontal table. Can a torch be so placed that the coin casts a circular shadow on the table? If so, what is the locus of possible

positions of the torch. And is the image of the hole ever in the centre of the circular shadow? If not, why not?

8. (i) Which triangles can one obtain as shadows of a given triangle? Can one perhaps get all triangles in this way?

(ii) Which triangles can one obtain as cross-sections of a given (infinitely extended) triangular pyramid? Can one obtain all triangles in this way?

9. (i) Which quadrilaterals can one obtain as shadows of a given quadrilateral? Can one get all quadrilaterals in this way?

(ii) Which quadrilaterals can one obtain as cross-sections of a given (infinitely extended) square pyramid?

10. How many edges must one slit in order to lay a polyhedron flat (in one piece)? In how many ways can this be done?

11. (i) You probably know a formula for the sum of the angles in an ordinary polygon with n sides. But what can you say about the sum of the angles at the five points of a pentagram like the one in Fig. 4.14(i)?

(ii) What if instead of having just five corners (as in the pentagram) a figure has n corners, with each corner joined to the next corner but one in each direction? What is the sum of the angles at the n corners of such a figure?

(iii) And what if instead of joining each corner to the corners 'two steps away' in each direction we join each corner to the corners 'three steps away' (as in Fig. 4.14(ii)). Can you find a formula for the sum of the angles at the n corners of such a figure?

(iv) Can you find a general formula for the sum of the angles in a figure with n corners, where each corner is joined to the corners 'k steps away' in each direction?

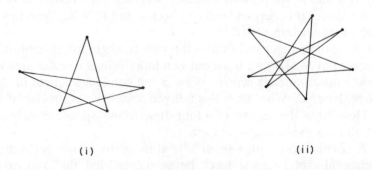

(i)   (ii)

Fig. 4.14

References for the further problems

3. H. S. M. Coxeter *Introduction to geometry*, pp. 211–212. Wiley, 1969.
5. H. Rademacher and O. Toeplitz *The enjoyment of mathematics*, pp. 163–177, Princeton University Press, 1970.
   R. Honsberger *Ingenuity in mathematics*, pp. 157–164, Mathematical Association of America, 1970.
6. A. Gardiner, *Infinite processes*, pp. 230–231, Springer, 1982.

4.4 Solutions to exercises in 4.2

Exercise 1 (i) There are many ways of doing this. For example, the line segment XY exactly halfway between the two parallel sides has length $(a + b)/2$. If you cut the trapezium in two along this line and then rotate the top half through $180°$ about the point Y, you get a parallelogram of height $\frac{1}{2}h$ with parallel sides of length $(a + b)$. This parallelogram, and hence the original trapezium, has area $\mathscr{A} = \frac{1}{2}(a + b) \times h$.

(ii) If $b = 0$ we get a triangle with area $\mathscr{A} = \frac{1}{2}a \times h$, so the same formula works. If $a = b = 0$ we get a mere line segment with area 0, so the same formula still works.

Exercise 2 A frustum of a square pyramid and a parallelepiped can both be constructed by starting with two parallel faces, and then joining them up in some way.

Exercise 3 (i) Your first attempt at a definition of a 'three-dimensional version of a trapezium' should probably try to incorporate the features mentioned in the solution to Exercise 2.

(iii) Your answer will depend on your definition. But the regular tetrahedron looks rather like a solid with two parallel faces—one of which has shrunk to a point. The cube is a special kind of parallelepiped. And what about the regular octahedron?

Exercise 4 Precisely when one of the two parallel edges of equal length is 'directly above the other'.

Exercise 5 (i) Volume = length × breadth × height. From this point of view a 'three-dimensional version of a rectangle' would presumably mean a *cuboid*.

(ii) Volume = (area of the two equal and parallel faces) × (perpendicular distance between them). From this point of view a 'three-dimensional version of a rectangle' is presumably a solid constructed by first taking two equal and parallel faces, one directly above the other, and then joining them up in some way.

Exercise 6 The hint shows how to cut up the given solid into (*a*) one $1 \times 1 \times h$ cuboid, (*b*) four corner tetrahedra, each of which has an $h \times \frac{1}{2}$ right-angled triangle as base and height $\frac{1}{2}$, and (*c*) four wedges each with a $1 \times \frac{1}{2}$ rectangular base and height h. So the total volume is $h + 4(h/24) + 4(h/4) = 13h/6 > A \times h$.

Exercise 7 (i) A right prism whose top and bottom faces are rectangles is just a cuboid. Since the area A of each rectangle is given by $A = \text{length} \times \text{breadth}$, the formula $\mathcal{V} = \text{length} \times \text{breadth} \times \text{height}$ for the volume of a cuboid can be written in the form $\mathcal{V} = A \times h$.

(ii) A carefully placed vertical cut produces two pieces which can be rearranged to give a right prism whose top and bottom faces are rectangles with the same area A as the original parallelograms. Since the rearranged solid has volume $A \times h$, so does the original solid.

(iii) Two identical copies of the given shape can be put together to make a right prism whose top and bottom faces are parallelograms, each having area $2A$. Since this latter solid has volume $2A \times h$, the original solid must have had volume $A \times h$.

(iv) A polygon with n sides can be cut into $n - 2$ triangles. In the same way one can make $n - 2$ vertical cuts to dissect a right prism whose top and bottom faces are polygons with n sides into $n - 2$ right prisms whose top and bottom faces are triangles. The sum of the areas $A_1, A_2, \ldots, A_{n-2}$ of these $n - 2$ triangles is just the area A of the original polygon with n sides. So the volume of the original solid is just $A_1 \times h + A_2 \times h + \ldots + A_{n-2} \times h = A \times h$.

(v) Approximate the circular base by inscribed regular polygons. Let A_n be the area of the regular 2^{n+1}-gon inscribed in the circular base, and let $\mathcal{V}_n$ be the volume of the corresponding right prism which has this regular 2^{n+1}-gon as base. Then $\mathcal{V}_n = A_n \times h$ by (iv). As n increases $A_n \to A$ and $\mathcal{V}_n \to \mathcal{V}$, so $\mathcal{V} = A \times h$.

Exercise 8 'A solid with two parallel faces which are joined together in some way.' (But how?)

Exercise 9 (c) is not a polyhedron because the 'faces' joining the top to the bottom are twisted instead of flat. (g) is a polyhedron, but is not a prismoid because the top and bottom faces are not joined directly corner to corner.

Exercise 10 (i) (a) $\mathcal{V} = A \times h = 1$. (b) $\mathcal{V} = \frac{1}{3}(A \times h) = \frac{1}{3}(\sqrt{3}/4 \times \sqrt{\frac{2}{3}}) = 1/6\sqrt{2}$. (d) This is a regular tetrahedron just like the one in (b), so $\mathcal{V} = 1/6\sqrt{2}$. (e) This regular octahedron can be cut into two square-based pyramids, each with volume $\frac{1}{3}(1 \times h')$, where $h' = 1/\sqrt{2}$, so $\mathcal{V} = \sqrt{2}/3$. (f) $\mathcal{V} = 13h/6$ by Exercise 6. (h) This shape can be cut up into a wedge ($=$ half a cuboid) with volume $\frac{1}{2}(1 \times h)$, and two rectangular-based pyramids each with volume $\frac{1}{3}(\frac{1}{2} \times h)$, so $\mathcal{V} = \frac{1}{2}h + \frac{1}{3}h = 5h/6$. (i) This shape can be cut up into simple pieces. But another method is to treat it as a truncated pyramid. Since the edge of the top face is exactly half the edge of the bottom face, the original pyramid must have had twice the height of the truncated pyramid, so $\mathcal{V} = \frac{1}{3}(4 \times 2h) - \frac{1}{3}(1 \times h) = 7h/3$.

(ii) (a) $A = 1$, $B = 1$, $h = 1$. (b) $A = \sqrt{3}/4$, $B = 0$, $h = \sqrt{\frac{2}{3}}$. (d) $A = 0$, $B = 0$, $h = 1/\sqrt{2}$. (e) $A = \sqrt{3}/4$, $B = \sqrt{3}/4$. $h = \sqrt{\frac{2}{3}}$. (f) $A = 2$, $B = 2$, h. (h) $A = 2$, $B = 0$, h. (i) $A = 4$, $B = 1$, h.

Exercise 11 (i) $\frac{1}{2}(A + B) \times h$.

(ii) It works for (a) but not for any of the others.

Exercise 12 The given frustum is obtained from a square-based pyramid of height $ha/(a - b)$ by chopping off the top $hb/(a - b)$. Hence $\mathcal{V} = \frac{1}{3}(a^2 \times ha/(a - b)) - \frac{1}{3}(b^2 \times hb/(a - b)) = \frac{1}{3}(a^2 + ab + b^2) \times h$.

Exercise 13 The hint gives one reason. You might also have noticed that a formula which refers only to A and B could never distinguish between the solid in Exercise 6 whose volume is $13h/6$, and the corresponding cuboid whose volme is $2h$.

Exercise 14 (i) Volume = (area of cross-section halfway up) × height.

(ii) No (because the cross-section halfway up is a square with side $(a+b)/2$. The formula works for any right prism so it works for (a); but it does not work for any of the other prismoids in Fig. 4.6.

Exercise   15 (i) $\mathscr{A} = (a+b)h/2 = (a+c+b)h/3 = (a+2c+b)h/4 = (a+3c+b)h/5 = (a+4c+b)h/6 = \ldots$.

(ii) In a right prism $A = B = C$, so $\mathscr{V} = (A+B)h/2 = (A+C+B)h/3 = \ldots$.

(iii) Suppose that two of these formulae both worked for some particular prismoid, say $\mathscr{V} = (A + (m-2)C + B)h/m = (A + (n-2)C + B)h/n$. Then $(A + (m-2)C + B)/m = (A + (n-2)C + B)/n$, so $A + B = 2C$. In other words the area of the cross-section halfway up is equal to the average of the areas of the top and bottom faces. This is clearly true in the case of a right prism (since then $A = B = C = (A+B)/2$), and so is true for (a) in Fig. 4.6; but it is false for the other six prismoids in Fig. 4.6.

(iv) For the frustum of a square pyramid in Exercise 12 we have $A = a^2$, $B = b^2$, $C = (a+b)^2/4$, and $\mathscr{V} = \frac{1}{3}(a^2 + ab + b^2)h$. We want to show that there is exactly one whole number m with the property that $(a^2 + ab + b^2)/3 = (a^2 + (m-2)(a+b)^2/4 + b^2)/m$ *for all possible values of a and b.* The previous part (iii) shows that this can be true for at most one value of m (since $C = (a+b)^2/4 \neq (a^2 + b^2)/2 = (A+B)/2$), and the equation is certainly true when $m = 6$; hence the equation is true for exactly one value of m. (*Alternatively*, if the equation is to hold for all possible values of a and b it must certainly hold when $a = 1$ and $b = 0$, so $\frac{1}{3} = (m+2)/4m$ and $m = 6$.)

This suggests that we should check whether the formula $\mathscr{V} = (A + 4C + B)/6$ holds for all other prismoids as well. It obviously works for all right prisms, because for them $A = B = C$ so the formula is equivalent to $\mathscr{V} = A \times h$. It also works for tetrahedra and pyramids, because for them $B = 0$, $C = A/4$, so the formula reduces to $\mathscr{V} = 2Ah/6 = \frac{1}{3}(A \times h)$. To check the other prismoids in Fig. 4.6 we must calculate the area C of the cross-section halfway between the two parallel faces. (*d*) The cross-section here is a regular hexagon with side $\frac{1}{2}$, so $C = \frac{1}{4}$. Since $A = B = 0$ we have $(A + 4C + B)h/6 = h/6 = 1/6\sqrt{2}$, which is precisely the volume we found in Exercise 10. (*e*) The cross-section here is a regular hexagon with side $\frac{1}{2}$, (so $C = 3\sqrt{3}/8$. Since $A = B = \sqrt{3}/4$, we get $(A + 4C + B)h/6 = 2\sqrt{3}h/6 = h/\sqrt{3}$, which is precisely the volume we found in Exercise 10. (*f*) The cross-section here is a square with side $\frac{3}{2}$, so $C = 9/4$. Since $A = B = 2$, we get $(A + 4C + B)h/6 = 13h/6$, which is precisely the volume we found in Exercise 6. (*h*) The cross-section here is a $1\frac{1}{2} \times \frac{1}{2}$ rectangle, so $C = \frac{3}{4}$. Since $A = 2$, $B = 0$, we get $(A + 4C + B)h/6 = 5h/6$, which is precisely the volume we found in Exercise 10.

Exercise 16 (i) If a prismoid has one vertex only, then the 'two parallel faces' E and F are both equal to this single point and $A = B = C = \mathscr{V} = 0$. If a

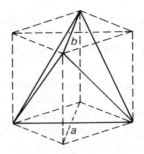

Fig. 4.15

prismoid has two vertices, then either (*a*) the 'two parallel faces' *E* and *F* are distinct and each consists of a single vertex, or (*b*) the 'two parallel faces' *E* and *F* are identical, (so $h = 0$) and each consists of the line segment joining the two vertices: in either case $A = B = C = \mathcal{V} = 0$. If a prismoid has three vertices, then either (*a*) the 'two parallel faces' *E* and *F* are distinct and one consists of a single vertex while the other consists of the edge joining the other two vertices, or (*b*) the 'two parallel faces' *E* and *F* are identical (so $h = 0$) and each is equal to the triangle spanned by the three vertices of the prismoid. In the first case $A = B = C = V = 0$, while in the second case $h = \mathcal{V} = 0$; so $\mathcal{V} = (A + 4C + B)h/6$ in each case.

(ii) It is easy to check that the formula works for the first kind of prismoid with four vertices—that is one in which one of the two parallel faces, say *E*, is a triangle and the other is a point: because then $B = 0$ and $C = A/4$, so the expression $(A + 4C + B)h/6$ reduces to the familiar $\frac{1}{3}(A \times h)$. The second kind of prismoid with four vertices—that is one in which the two parallel faces *E*, *F* are both line segments of length *a* and *b* respectively—is more awkward. The areas *A* and *B* of *E* and *F* are obviously equal to zero, and the cross-section halfway between *E* and *F* is a parallelogram (why?) with sides of length $\frac{1}{2}a$ and $\frac{1}{2}b$, the angle between these two sides being precisely the angle between the directions of the line segments *E* and *F*. Perhaps the easiest way to find the volume of such a prismoid *in terms of A, B, C, and h* is to first apply a horizontal shear so that the midpoint of the edge *F* is directly over the midpoint of the edge *E*. This transformation does not affect the volume $\mathcal{V}$, the height *h*, or the area *C* of the cross-section halfway up. (Why not?) One can then imagine the prismoid as sitting inside a right prism whose base is a parallelogram with diagonals of length *a* and *b* (Fig. 4.15). The area of the base is exactly $2C$ so the volume of the right prism is equal to $2C \times h$. The prismoid is obtained by cutting away four tetrahedra of total volume $4Ch/3$, so the original prismoid has volume $2Ch/3 = (A + 4C + B)h/6$.

(iii) The hint should allow you to see this on your own.

Part 2

Extended investigations

Part 2

Part 2

Extended investigations

Advice to the reader

If it were easy, the book ought to be burned, for
it cannot be educational. In education, as else-
where, the broad primose path leads to a nasty place.
(A. N. Whitehead, *The organisation of thought*)

Part 2 contains two *extended* investigations in elementary mathematics. They may be tackled in either order. The second investigation (The Postage Stamp Problem) has the advantage that the initial conjecture can be formulated on the basis of very simple numerical work; but this does not get one very far and other ideas soon have to be introduced. The first investigation (Flips) does not begin in quite such a simple-minded way, but it has the advantage that the initial naive approach can be sustained throughout; and though it too has to be supplemented eventually by more powerful methods, it is never actually supplanted by them.

Each investigation focusses on a single elementary problem, which is then explored in a sequence of **exercises** interspersed with **text**. The text and exercises have been carefully structured to guide you in your exploration of the initial problem and its natural ramifications. They map out one way of exploring that initial problem. If you follow them through and use the hints and solutions as I suggest, then you should manage to solve the central problem in each investigation. But the path I have outlined is offered as a guide, not a straitjacket. I shall have succeeded only if you use the text as a starting point and begin to ask your own questions and to explore your own ideas. At the end of Part 2 you will find a collection of Further problems which are well worth investigating in the same spirit as 'Flips' and 'The postage stamp problem'.

The central problem in each investigation has been chosen with the following five criteria in mind.

(1) The problem should be easy to understand, and must be sufficiently interesting for the reader to want to solve it.

(2) The solution should not be obvious.

(3) The attempt to solve it should lead to some interesting mathematics.

(4) The mathematical techniques required to solve the original problem should be entirely elementary.

(5) And last, but most important of all, the whole process—from formulating the initial problem right through to the final solution—should suggest the kind of approach a mathematician would use to tackle the same problem if he knew no more than secondary school mathematics.

You have two separate things to do.

I. Do all the exercises.

II. Read the text.

I. The exercises are like mathematical *bricks*. They provide the basic experiences you need in order to discover some of the interesting questions which lie behind the original problem, and to piece together your own answers to these questions.

First warning You may be tempted to dash through the exercises giving them very little thought, as if you were working through a school textbook and the main aim was to get to the end as quickly as possible. **Slow down**. Think all the time about what you are doing and why you are doing it. Before beginning an exercise think carefully about what you have been asked to do, and try to see whether the exercise is intended to help you to discover something new, or simply to check what you thought you knew already. And when you have completed an exercise ask yourself whether your answer fits in with what you should have expected.

Each exercise has been designed to help you think about some particular aspect of the main problem in the hope that you will gradually construct your own solution. If you rush on without thinking you will miss the satisfaction of working things out for yourself, and you will sooner or later discover that you no longer understand what is going on.

II. The text is like the *mortar* which cements the exercises together.

Second warning You may be tempted to skate over the written text and hurry on to the next exercise. **Slow down**. Read the text carefully. Sometimes the text is used to give a summary of progress so far: when it does so check that the summary agrees with what you found. At other times the text is used to introduce a new idea: you should then make sure you understand the idea before charging on to the next exercise. Occasionally the text combines a review of where we have got to with a discussion of why we are about to set off in in a slightly unexpected direction: the exercises which follow such a change of direction will not make much sense if you have not thought about the motivation behind them.

There is relatively little text. But if you do not read it carefully you are likely to lose your way, and will land up doing (or rather trying to do) what you are told without understanding why it is a sensible thing to do in the circumstances.

When working on your own you are bound to get stuck at times. Many of the exercises have hints: these are printed at the end of each chapter, after the text and before the solutions.

Third warning   In mathematics there are often many different ways of tackling the same exercise. Your aim should be to find your own solutions. And your solutions may be different from mine. If you are still in the middle of trying to work out your solution, then my hint might cause you to abandon your own ideas too soon. But the hints and solutions are there to help you, so don't be afraid of using them when you need them.

Some exercises have been marked with an asterisk. These exercises and their solutions are particularly important if you are to understand later work. If you fail to solve such an exercise even after consulting the appropriate hint, then the obvious thing to do is to try to understand my solution before going any further. An equally sensible strategy if you fail to solve an exercise with an asterisk is to go on to the next piece of text and have a go at the next few exercises keeping an eye open for new ideas which might help you solve the offending exercise on your own. But when you come to a 'grey line' marking a break in the text, like this

---

it is time to go back and make sure that you have mastered all those exercises which were marked with an asterisk, if necessary by consulting my solutions.

For many of you this will be your first encounter with an extended piece of mathematics. It may take a couple of hours before you begin to appreciate what is going on, and you may have to push yourself at first. When you eventually begin to uncover the unexpected mathematics behind the original problem you will appreciate that the initial effort was well worthwhile. But though discipline and stamina are required, the work is supposed to be both challenging *and enjoyable*. So if you get bogged down on a particular exercise, put it to one side and read on: the subsequent text and exercises will usually sort out most of your difficulties.

Finally, these investigations are not 'all-or-nothing' affairs. You will obviously miss something if you only complete part of an investigation. But I have tried to design each one in such a way that no matter where you stop, you will still have gained something worthwhile from the experience. So **if at any stage you feel you have had enough, don't be afraid to stop in the middle of an investigation.** You may miss something, but I think you will find that you have still gained a great deal.

Investigation I

Flips

5 Introducing 9-flips

*'It seems very pretty,' she said when she had finished
it, 'but it's rather hard to understand!' (You see she
didn't like to confess, even to herself, that she
couldn't make it out at all.)*

(Alice in *Through the looking-glass*)

In Chapter 2 we investigated a bit of 'number magic'. If you did not
work through the following exercise then, you should do so now.

Exercise 1 (i) Pick a three-digit number *abc* whose first and last
digits are different.

$$a \ b \ c$$

Reverse the order of the digits to get another three-digit number.

$$c \ b \ a$$

Subtract the smaller of the two numbers from the larger one to get a
new three-digit number *def*. (If you get a two-digit number *ef*, write it
as a three digit number 0*ef* with first digit zero.)

$$
\begin{array}{c}
a \ b \ c \\
\underline{c \ b \ a} \\
d \ e \ f
\end{array}^{-}
$$

Add your new three-digit number *def* and its reverse *fed*. (If your new
three-digit number 0*ef* has first digit zero, reversing it will give a
three-digit number *fe*0 with last digit zero.) *What answer do you get?*

$$
\begin{array}{c}
d \ e \ f \\
\underline{f \ e \ d} \\
\cdot \ \cdot \ \cdot \ \cdot
\end{array}^{+}
$$

(ii) Pick another three-digit number *abc* and repeat the same
process. What answer to you get this time?

(iii) Get a friend to pick a secret three-digit number. Give the same
sequence of instructions you followed in (i), except that the answer

should be kept secret. Then pretend to do some calculations of your own before 'discovering' the answer.

(iv) Now try to figure out why you always get the same answer 1089.

This investigation starts out from another intriguing property of the number $1089_{\text{base 10}}$.

Exercise 2* (i) Work out 1089×9. What do you notice about the answer?

(ii) Try to find some other four-digit numbers which behave like 1089 when multiplied by 9.

Suppose that N is a positive number written base 10, and that $9 \times N$ has the same digits as N but in reverse order. Then we shall say for short that N is a **9-flip**.

In Exercise 2(i) you showed that 1089 is a 9-flip with four digits. In Exercise 2(ii) you were challenged to find some other 9-flips with four digits. This is quite hard as you probably discovered. What if you couldn't find any? Could it perhaps be true that 1089 is the only 9-flip with four digits? If so, is there any hope of finding all 9-flips no matter how many digits they have?

But before you rush off and try to find first all 9-flips with five digits, then all 9-flips with six digits, and so on, it might be worth taking a step backwards to look for all 9-flips with one digit, with two digits, and with three digits.

Exercise 3* Are there any 9-flips with one digit?

Exercise 4* Are there any 9-flips with two digits?

Exercise 5* Find all 9-flips with three digits?

One way of answering Exercise 4 would be to work out 9×10, 9×11, 9×12, and so on all the way up to 9×99. In other words you could simply *test each two-digit number in turn* to see if it is a 9-flip. This would certainly work. But the method is too simple-minded to be really useful. What you want is a simple but effective method of finding not just all 9-flips with five-digits, but 9-flips with fifty digits and 9-flips with five hundred digits too! Have a look at the hints (and if necessary my solutions) to Exercise 4 and Exercise 5 before going any further.

You should now be in a position to improve on your solution to Exercise 2(ii) by filling in the details of the next exercise.

Exercise 6 Let *a*, *b*, *c*, *d* be the digits of a 9-flip *abcd*_{base 10}. Use the hints on the right to answer the following questions.

(i) What must the digit *a* in the thousands column be? So what must the digit *d* be?

How many ten thousands are carried over from the thousands column?

(ii) You now know the digits *a* and *d*. How many thousands are carried over from the hundreds column? What are the two possible values for *b*?

You know that a = 1, d = 9, and that nothing is carried from the hundreds to the thousands column.

(iii) Show that one of these two possible values for *b* cannot happen.

If b = 1, then c = ___, and . . .

(iv) Now that you know what *a*, *b* and *d* must be, find all possible values of *c*. Hence find all 9-flips with four digits.

You should now try to use the same method to find all 9-flips with five digits.

Exercise 7 Let *a*, *b*, *c*, *d*, *e* be the digits of a 9-flip *abcde*_{base 10}.

(i) What must the digit *a* be? What must the digit *e* be?

(ii) What are the two possible values for *b*? Show that one of these values cannot happen.

(iii) Find all 9-flips with five digits.

You now know that there are no 9-flips with one, two, or three digits, and that there is just one 9-flip with four digits. And if you completed Exercise 7 successfully you also know the complete list of 9-flips with five digits. **But where is the mathematics in all this?**

It is often a good idea to work your way into a problem by doing some initial calculations. But sooner or later you have to decide whether to plough on with more and more calculations, or to stop calculating for a moment and think instead about where you should be going. Now that you have done some initial calculations with 9-flips you should perhaps try to to decide which questions look as if they are worth investigating. For example,

Question A *Is it possible to say exactly how many 9-flips there are with precisely n digits?*

Question B *Is it possible to describe all 9-flips, or all 9-flips with precisely n-digits, in some simple way?*

You have in some sense already begun to answer Questions A and B. In Exercises 3–7 you found all 9-flips with at most five digits. And it is all too easy to suggest, without thinking, that the next thing to do is to find all 9-flips with six digits, with seven digits, and so on.

Exercise 8* (Just in case you never completed Exercise 7) Find all 9-flips with five digits.

Exercise 9* Use the method outlined in Exercise 6 to find all 9-flips with six digits.

   The answers to Exercises 8 and 9 will be very useful later on. But there are two things wrong with the thoughtless way in which we went straight on from numbers with four digits to numbers with five digits, and from numbers with five digits to numbers with six digits.
   First, you may answer Question A and Question B completely for numbers with six, seven, eight, nine and even ten digits, but this may not help you answer either Question A or Question B for numbers with a hundred, or a million digits.
   Second, when you answered Question A and Question B for numbers with five digits you will have noticed that you were simply repeating a lot of the work you did for numbers with four digits (Fig. 5.1). So when you plough through the detailed calculations for numbers with four, five, six, seven, or eight digits, you should not be

$$
\begin{array}{r}
\overset{1}{a}\ b\ c\ d\ \overset{9}{e} \times \\
9 \\
\hline
\underset{9}{e}\ d\ c\ b\ \underset{1}{a} \\
\hline
\end{array}
$$

Hint

$a = 1$, so $e = 9$

$\therefore b = 0$ or $b = 1$

Etc.

Fig. 5.1

content merely to 'get the right answers'. You must also be permanently on the lookout for clues which may help you discover general patterns.

But Question A and Question B are only a beginning. Before we could claim to have mastered 'the mathematics of 9-flips' we would like to know the answers to lots of other interesting questions. For example

Question C *What about 8-flips, 7-flips, and so on? Can you find them all?*

Question D *What happens if the numbers that you work with are written in other bases? For example, if you write all numbers in base 9, are there any 8-flips at all? If so, can you find them all?*

Question E *Is there some connection between Exercise 1 and Exercise 2? Why does the same number crop up in both places?*

But to begin with we shall stick with numbers written in base 10 and concentrate our efforts on trying to answer Question A and Question B.

Hints to Exercises

2. (i) Read your answer backwards.
   (ii) Suppose a, b, c, d are digits between 0 and 9 with $a \neq 0$.

What can you say about a? What can you say about d?

4. $9 \times a$ gives a single digit with no carrying.
   $\therefore a = $ _____ and $b = $ _____ .

5. $9 \times a$ gives a single digit with no carrying.
   $\therefore a = $ _____ and $c = $ _____ .

Solutions

Exercise 1 See the solution to Exercise 3 in Chapter 2.

Exercise 2 (i) The digits of 1089 are simply reversed to give 9801.

Exercise 3 It all depends whether or not you count '0' as a one digit number.

Exercise 4 No. (The hint shows that if ab is a 9-flip with two digits, then $a = 1$ and $b = 9$. But $9 \times 19 \neq 91$, so 19 is not in fact a 9-flip.)

Exercise 5 There aren't any. (The hint shows that if abc is a 9-flip with three digits, then $a = 1$ and $c = 9$. But there can be no carrying from the tens to the hundreds column, so $b = 0$ or $b = 1$. However $9 \times 109 \neq 901$ and $9 \times 119 \neq 911$, so neither 109 nor 119 is in fact a 9-flip.)

Exercise 6 (i) $a = 1$, $d = 9$.

(ii) None. Hence $b = 0$ or $b = 1$.

(iii) If $b = 1$, then $9 \times c + 8$ must end in a 1. So $9 \times c$ ends in a 3. Therefore $c = 7$. But $9 \times 1179 \neq 9711$, so b cannot equal 1.

(iv) Hence $b = 0$ and $9 \times c + 8$ must end in a 0. So $9 \times c$ ends in a 2. Therefore $c = 8$. Hence $abcd = 1089$ is the only 9-flip with four digits.

Exercise 7 (i) $a = 1$, $e = 9$.

(ii) $b = 0$ or $b = 1$. If $b = 1$, then $9 \times d + 8$ must end in a 1. So $9 \times d$ ends in a 3. Therefore $d = 7$ and 7 has to be carried to the hundreds column; but then $9 \times c + 7$ must end in c, which is impossible, so b cannot equal 1. (*Alternatively*, if $b = 1$, then nothing is carried over from the thousands column, so $d = 9 \times b + ? = 9 + ? = 9$. But from the tens column we see that $9 \times d + 8$ is supposed to end in $b = 1$.)

(iii) $a = 1$, $c = 9$, $b = 0$. Then from the tens column $9 \times d + 8$ must end in 0. So $d = 8$, and 8 has to be carried over to the hundreds column. But then $9 \times c + 8$ must end in c and must carry 8 to the thousands column, so $c = 9$. hence $abcde = 10989$. And since $9 \times 10989 = 98901$, there is precisely one 9-flip with five digits.

Exercise 8 See the solution to Exercise 7(iii).

Exercise 9 Let $abcdef$ be a 9-flip with six digits. Then $a = 1$, $f = 9$, $b = 0$, $c = 8$ as in Exercise 7. In the thousands column $9 \times c + ?$ must carry 8 to the ten thousands column, so either $c = 8$ (and $? = 8$, so $d = 0$) or $c = 9$. But if $d = 0$, then nothing is carried to the thousands column, so $? = 0$. Therefore $c = 9$, and $9 \times d + 8$ ends in 9, so $d = 9$. Hence $abcdef = 109989$. And since $9 \times 109989 = 989901$, there is precisely one 9-flip with six digits.

6 The art of guessing

It is much easier to detect error than to reach truth.
(Galileo)

Take another look at Question *B*.

Question *B* *Is it possible to describe all 9-flips in some simple way?*

An optimist might prefer to replace this slightly tentative question by a straight challenge.

Exercise *B* Find all 9-flips.

At first sight Exercise *B* looks very similar to several problems you have already solved. For example

Exercise 6 (iv) Find all 9-flips with four digits.

Exercise 7 (iii) Find all 9-flips with five digits.

Exercise 9 Find all 9-flips with six digits.

But there is one big difference between Exercise *B* and Exercises 6, 7, and 9. In Exercise 6 you considered an unknown 9-flip *abcd*, and showed
first, that $a = 1$ and $d = 9$;
second, that $b = 0$ or $b = 1$;
third, that $b = 1$ cannot happen, so $b = 0$;
and finally, that $c = 8$, so $abcd = 1089$.
For 9-flips with five digits and 9-flips with six digits the calculation is a bit longer, but it is still not too hard to write out a complete answer. But in Exercise *B* a complete answer would have to tell us not only about 9-flips with four digits, five digits, and six digits, but also about 9-flips with four hundred digits, five thousand digits, and six million digits. So you should not really expect to solve Exercise *B* in quite the same way as you answered Exercise 6.
The difference between Exercise 6 and Exercise *B* is partly due to

the fact that Exercise 6 presents us with a *finite* task, whereas Exercise *B* looks like an *infinite* task. To find all 9-flips *with four digits* you could try looking at each of the numbers from 1000 to 9999 one at a time, multiplying each one by 9 to see whether the number you started with is a 9-flip. This may not be a very satisfactory way of finding all 9-flips with four digits, but it would certainly work. The approach outlined in Exercise 6 is in some sense just a clever way of checking all the numbers between 1000 and 9999. But if we want to find *all* 9-flips, and not just those with four digits, then we have to find some very clever way of checking *every positive whole number*!

How would a mathematician begin to tackle such a problem? When the answer to the problem is not obvious, as in the case of Exercise *B*, it is often a good idea not to try to tackle the problem head-on, but to try sneaking up on it one step at a time.

If you were sneaking up on someone in real life you would only need to use two different steps: 'left foot forward' and 'right foot forward'. But by repeating these two steps over and over again, one after the other, you could go as far as you needed to. The way we shall sneak up on Exercise *B* is also based on just two basic steps, which we shall simply repeat over and over again until we finally succeed in tracking down what looks like a complete answer to Question *B*. The two steps we shall use over and over again in our search for a solution to Exercise *B* are

Exercise B: Step 1 *Guess*! Use any methods you like to guess what looks like a complete list of 9-flips.

Exercise B: Step 2 *Check your guess*! Test your guess in some way to find out what, if anything, is wrong with it.

When you try to repeat these two steps over and over again you will find yourself doing something like this.

Step 1 Make an intelligent *first guess* at what looks like a complete list of 9-flips.

Step 2 Try to find some new 9-flips which you missed out of your first list.

Step 1 again Improve your first guess to take account of the new 9-flips you discovered in Step 2.

Step 2 again Try to find some new 9-flips which you missed out of your (improved) second list.

Step 1 again . . .

What is the point of deliberately avoiding a head-on attack on Exercise *B* and choosing instead to repeat these two steps over and over again?

This 'Guess–Check–Guess again–Check again– . . .' procedure is certainly not enough to provide a complete answer to Exercise *B*. *But it is an excellent way of getting started.* At the very least it gives you a chance of coming to grips with a problem which at first sight you had no idea how to solve. As we said before, it is a way of edging your way, little by little, towards a complete answer.

When you eventually think you are absolutely certain what the complete list of 9-flips must look like, it will be time to worry about

Exercise B: Step 3 *Prove*! Give a convincing mathematical reason, or proof, which shows that your list really is complete.

But the whole purpose of the 'Guess–Check–Guess again–Check again– . . .' procedure is to put off worrying about Step 3 until you have played around with the problem for a while and have come up with something which really seems to be true, and hence worth trying to prove.

It is time to see how all this is going to help you investigate Question *B*. So let's start with Step 1.

Step 1 *Guess*! Make an intelligent first guess at what looks like a complete list of 9-flips.

In the previous chapter you found all 9-flips with at most six digits. The table in Fig. 6.1 presents a summary of what you already know. It does not require much imagination to make a *first guess* at how this table goes on.

Number of digits	Complete list of 9-flips
1	None (Exercise 3)
2	None (Exercise 4)
3	None (Exercise 5)
4	1089 (Exercise 6)
5	10989 (Exercises 7, 8)
6	109989 (Exercise 9)
7	?
.	.

Fig. 6.1

Exercise 10 What do you expect the next few entries in the table to be? Fill in your guesses in the table of Fig. 6.2.

Number of digits	Your *guessed* list of all possible 9-flips
7	?
8	?
9	?
10	?
11	?

Fig. 6.2

Though you filled in your *first guess* of the complete list of 9-flips only for 9-flips with seven, eight, nine, ten, and eleven digits you probably have a fairly clear idea what the complete list of 9-flips looks like. But you should not forget that this idea is *only a first guess*. You must therefore proceed to Step 2.

Step 2 *Check your guess*! Test your guess in some way to find out what, if anything, is wrong with it.

There are in fact two different things to check about your guessed list of all possible 9-flips.

Exercise 11 First you should check each of the numbers you wrote in the table in Exercise 10 to see whether it really is a 9-flip. Do this now!

If some of the numbers you wrote down turned out not to be 9-flips, you would have to go straight back and look for a better guess (Step 1 again). But what if all the numbers in your guessed list really are 9-flips? Does it follow that your guess was completely correct? You may be tempted to think that it does. *But you would be wrong!* For you have not yet carried out the second, and most important, check on your guessed list of 9-flips.

Step 2 *Check your guess! How do you know that your guessed list contains all possible 9-flips? How do you know that you have not missed some out?*

There is no foolproof way of deciding whether or not you have missed out some 9-flips from your list. But you can at least begin by checking whether your guessed list includes all possible 9-flips with (say) seven, eight, or nine digits.

Exercise 12* (i) Use the method outlined in Exercise 6 to find all 9-flips with seven digits.

(ii) Now look back at your *guessed* list of 9-flips with seven digits in Exercise 10. Compare your guessed list with the complete list you found in Part (i) above. Was your guessed list complete?

Exercise 13* (i) Use the method outlined in Exercise 6 to find all 9-flips with eight digits.

(ii) Compare this complete list of 9-flips with eight digits with the *guessed* list you entered in the table in Fig. 6.2. Was your guessed list really complete?

Exercise 14* (i) Find all 9-flips with nine digits.

(ii) Was the *guessed* list of 9-flips with nine digits which you entered in the table of Fig. 6.2 really complete?

The purpose of guessing is not just to be right the first time. If the problem you are trying to solve is at all difficult, then your first guess is almost bound to be wrong. But don't let this stop you trying to guess as intelligently as you can. The real value of guessing is that it makes it possible to replace a general problem which may be much too difficult to solve all at once, such as

Exercise B *Find all 9-flips*

by a whole family of much more manageable, and much more specific questions, such as

Exercise B: Step 2 *Is my guessed list of 9-flips with seven digits complete? Is my guessed list of 9-flips with eight digits complete? Is my guessed list of 9-flips with nine digits complete?* ...

An intelligent guess acts like a lens by focussing one's attention on something *specific*. And it is almost always easier to think about something specific, like Exercises 12, 13, and 14, than it is to think about something general, like Exercise *B*.

To end this chapter you should have a go at trying to answer Exercise *B* for yourself. But we shall come back to the same problem later on, so don't be too worried if you get stuck.

Your answers to Exercises 13 and 14 should have convinced you of the need to improve on your first guess. For example, when you used the method of Exercise 6 to find all 9-flips with eight digits you presumably started like this.

$$\begin{array}{r} abcdefgh \\ \times\ 9 \\ \hline hgfedcba \end{array}$$

It is easy to see
   first, that $a = 1$ and $h = 9$;
   second, that $b = 0$ or $b = 1$;
   third, that $b = 1$ cannot happen, so $b = 0$;
   fourth, that $g = 8$, so you have to carry 8 from the
      hundred thousands column to the millions column;
   hence $c = 8$ or $c = 9$.
If $c = 9$ you get the 9-flip you expected. But if $c = 8$ you get a 9-flip which you probably overlooked when you made your first guess in Exercise 10. As soon as you became aware of this unexpected 'intruder' you should have tried to improve on your *first* guess by coming up with a *second* guess which took account of what you had just discovered. Your answer to Exercise 14 would probably then have shown that your second guess was still not right. You would then have had to improve your guess once more.

See whether you can carry on this process of 'guessing and checking' until you really know the answer to Question *B*, even if you still can't see at this stage how to prove that your guess is correct (Step 3).

Exercise 15* (i) Use your answers to Exercises 13 and 14 to fill in the complete list of 9-flips with eight digits and 9-flips with nine digits in the table in Fig. 6.3.

Number of digits	Complete list of 9-flips
4	1089
5	10989
6	109989
7	1099989 (Exercise 12)
8	(Exercise 13)
9	(Exercise 14)

Fig. 6.3

(ii) What do you expect the next few entries to be? Fill in your guesses in the table in Fig. 6.4.

Number of digits	Your *guessed* list of all possible 9-flips
10	
11	
12	
13	
.	

Fig. 6.4

(iii) Now use the method of Exercise 6 and check your guessed list of 9-flips with ten digits to see whether you have missed any out. Then check your guessed lists of 9-flips with eleven digits, with twelve digits, and with thirteen digits in the same way.

(iv) Carry on 'guessing' (Step 1) and 'checking your guesses' (Step 2) until you are certain that you have discovered a really simple '*recipe*' which tells you how to write down *all* 9-flips with (say) twenty four digits.

(v) Try to find a convincing mathematical explanation why your 'recipe' does not miss out any 9-flips.

Hints to exercises

14. (i) Use the method outlined in Exercise 6 to find them all.
15. (v) Use the same method which worked in Exercises 6, 8, 9. But this time start with a 9-flip

$$abcd \ldots wxyz$$

which has an *unspecified number of digits*. First try to pin down the 'outside' digits a and z. Then work your way steadily inwards.

$$\overset{1}{a}bcd \ldots wxy\overset{9}{z} \times \\ \underset{9}{9} \\ \overline{\underset{9}{z}yxw \ldots dcb\overset{9}{a}\underset{1}{}}$$

Solutions

Exercise 12 (i) You should by now be able to extend the reasoning of Exercises 6, 7, and 9 to show that 1099989 is the only 9-flip with seven digits.

Exercise 13   (i) Let *abcdefgh* be a 9-flip with eight digits. Then as usual $a = 1$, $h = 9$, $b = 0$, $g = 8$. Then $9 \times c + ?$ must carry 8 to the millions column, so either $c = 8$ (and $? = 8$, so $f = 0$), or $c = 9$. If $c = 9$, then $f = d = e = 9$ and you get the expected 9-flip 10999989. If $c = 8$, then $f = 0$, $d = 9$, $e = 1$ and you get a 9-flip which you may not have expected, namely 10891089.

Exercise 14   (i) Let *abcdefghi* be a 9-flip with nine digits. Then $a = 1$, $i = 9$, $b = 0$, $h = 8$ as usual. $9 \times c + ?$ must carry 8 to the next column so either $c = 8$ (and $? = 8$ so $g = 0$), or $c = 9$. If $c = 9$, then $g = d = f = e = 9$ and you get the expected 9-flip 109999989. If $c = 8$, then $g = 0$, $d = 9$, $f = 1$, $e = 0$ and you get a 9-flip which you may not have expected, namely 108901089.

Exercise 15   (iv), (v) I shall not give the solution at this stage. Keep chewing the problem over for a while. We shall come back to it again later (for example, in Exercises 24 and 30).

7 Other flips

*The scientist [...] is on the look-out for events which are not yet **quite** intelligible, but which could probably be mastered as a result of some intellectual step which he has the power to take.*

(Stephen Toulmin, *Foresight and understanding*)

You might well ask why we started with 9-flips rather than 7-flips, 4-flips, or some other flips. One of the questions we asked at the end of Chapter 5 raised precisely this point.

Question *C* *What about 8-flips, 7-flips, and so on? Can you find them all?*

Now that you have had a really good go at finding all 9-flips, it is worth trying to use exactly the same methods to track down all other flips. Don't be put off by the fact that this looks like a very long job, for you have in fact done all the hard work in Chapters 5 and 6. You will probably find that you can get through this short chapter surprisingly quickly.

A 1-flip is just a number like 123454321 which reads exactly the same both forwards and backwards. Such numbers are called *palindromes*. They occur unexpectedly in many parts of mathematics, but we shall not say any more about them here. So the only things left for you to do in this chapter are to find all 8-flips, all 7-flips, all 6-flips, all 5-flips, all 4-flips, all 3-flips, and all 2-flips!

Exercise 16* Find all 8-flips with four digits.

Once you have answered Exercise 16 you should have no difficulty finding all 8-flips.

Exercise 17* Find all 8-flips.

And once you have succeeded in finding all 8-flips you should be able to use exactly the same method to find all 6-flips.

Exercise 18* (i) Find all 6-flips with four digits.
(ii) Find all 6-flips.

By making use of similar ideas you should be able to find all 7-flips, all 5-flips, and even all 4-flips, all 3-flips, and all 2-flips.

Exercise 19 (i) Find all 7-flips with four digits.
(ii) Find all 7-flips.

Exercise 20 (i) Find all 5-flips with four digits.
(ii) Find all 5-flips.

Exercise 21* Find all 4-flips with four digits.

You are obviously going to have to think a bit harder before trying to find all 4-flips. But the work you did in Chapters 5 and 6 when trying to find all 9-flips should give you a fairly clear idea about how to proceed.

Exercise 22 (i) Find all 4-flips with one digit.
(ii) Find all 4-flips with two digits.
(iii) Find all 4-flips with three digits.

Exercise 23* Find all 4-flips with five digits.

Once you have answered Exercises 21, 22 and 23 you should feel that you are on familiar ground. You should now be in a position to make an intelligent first guess about 4-flips with six, seven, eight, nine and ten digits.

Exercise 24* (i) Use your answers to Exercises 21, 22, and 23 to complete the table in Fig. 7.1.

Number of digits	Complete list of 4-flips
1	
2	
3	
4	
5	

Fig. 7.1

(ii) What do you expect the next few entries in the table to be? Fill in your guesses in the table in Fig. 7.2.

Number of digits	Your *guessed* list of all possible 4-flips
6	
7	
8	
9	
10	

Fig. 7.2

(iii) Now check each of your guesses.

(iv) Give a very simple 'recipe' which tells you how to write down all 4-flips with (say) thirty four digits.

(v) Find a convincing mathematical explanation of why your 'recipe' does not miss out any 4-flips.

There is a slightly surprising connection between 9-flips and 4-flips which you may not have noticed already. This is the subject of the next Exercise.

Exercise 25 (i) There is just one 9-flip with four digits (1089), and just one 4-flip with four digits (2178). How are these two numbers related?

(ii) There is just one 9-flip with five digits (10989), and just one 4-flip with four digits (21978). How are these two numbers related?

(iii) Write down the two 9-flips with eight digits and the two 4-flips with eight digits. How are these numbers related to one another?

(iv) Can you explain why 9-flips and 4-flips are related in this way?

To round off this chapter on 'Other flips' you should finish the job by considering 3-flips and 2-flips.

Exercise 26 (i) Find all 3-flips with four digits.
(ii) Find all 3-flips.

Exercise 27 (i) Find all 2-flips with four digits.
(ii) Find all 2-flips.

Hints to exercises

16. You could use the method outlined in Exercise 6. But there is a much quicker way. What does the *units column* tell you about *a*?

17. Read the hint for Exercise 16.
18. (i) Replace '8' by '6' in the hint for Exercise 16.
19. (i) What can you say about *a*? What does the units column tell you about *d*?

$$\begin{array}{r} a\ b\ c\ d \\ 7 \times \\ \hline d\ c\ b\ a \end{array}$$

(ii) Read the hint for Part (i) again.
20. (i) Replace '7' by '5' in the hint for Exercise 19.
21. From the units column you know that *a* is the units digit of $d \times 4$, so *a* must be *even*. From the thousands column you can see that $a = 1$ or $a = 2$.
22. (ii) Look again at the hint to Exercise 4.

(iii) Look again at the hint to Exercise 5.
24. (v) When you are thinking about 4-flips, bear in mind what you discovered about 9-flips in Chapter 5 and Chapter 6.

Solutions

Exercise 16 Let *abcd* be an 8-flip with four digits. Then $8 \times d$ ends in *a*, so *a* must be even. But then $a \geqslant 2$, so $8 \times a + ?$ would spill over into the ten thousands column. So there are no 8-flips with four digits.

Exercise 17 Exactly the same reasoning as in the solution to Exercise 16 shows that there are no 8-flips with *n* digits, no matter how large *n* may be.

Exercise 18 (i) Replace '8' by '6' in the solution to Exercise 16.
(ii) Replace '8' by '6' in the solution to Exercise 17.

Exercise 19 (i) Let *abcd* be a 7-flip with four digits. Then $a = 1$ as usual. But then $7 \times d$ must end in a 1, so $d = 3$. However $7 \times 1bc3$ must be greater than 7000 and so cannot equal $3bc1$.
(ii) Exactly the same reasoning as in (i) shows that there are no 7-flips at all.

Exercise 20 (i) Let *abcd* be a 5-flip with four digits. Then $a = 1$ as usual, so $5 \times d$ must end in a 1 which is impossible.
(ii) Exactly the same reasoning as in (i) shows that there are no 5-flips at all.

Exercise 21 The hint shows that $a = 2$. Then $4 \times 2 + ? = d$, so either $d = 8$ (and $? = 0$, so $b \le 2$), or $d = 9$. But $4 \times d$ ends in 2, so $d = 3$ or $d = 8$. Hence $d = 8$. Then $4 \times c + 3$ ends in b, so b must be odd. Hence $b = 1$, and $c = 2$ or $c = 7$. But $c = 2$ makes the hundreds column go wrong, so $c = 7$. Hence $abcd = 2178$. And since $4 \times 2178 = 8712$, there is precisely one 4-flip with four digits.

Exercise 22 (i) It all depends on whether or not you count '0' as a single digit number.

(ii) There aren't any. (The hint shows that if ab is a 4-flip with two digits, then $a = 2$ and $b = 8$. But $4 \times 28 \ne 82$, so 28 is not in fact a 4-flip.)

(iii) There aren't any. (The hint shows that if abc is a 4-flip with three digits, then $a = 2$ and $c = 8$. So $4 \times b + 3$ must end in b. Hence $b = 9$. But this is impossible since nothing is supposed to be carried to the hundreds column.)

Exercise 23 Let $abcde$ be a 4-flip with five digits. Then $a = 2$, $e = 8$ as usual. And $b = 1$ (as in the solution to Exercise 21), so $d = 7$. But then $4 \times c + 3$ must end in c, so $c = 9$. Hence $abcde = 21978$. And since $4 \times 21978 = 87912$, there is precisely one 4-flip with five digits.

Exercise 24 (iv) The basic 4-flips are 0, 2178, 21978, 219978, 2199978, ... These can be strung together in any way, as long as the first and last blocks, the second and second last blocks, and so on, are identical in pairs. For example

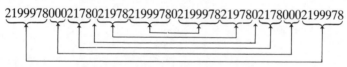

$$21999780002178021978219997802199978219780217800021 99978$$

(v) We shall come back to the reason **why** all 4-flips (base 10) are obtained in this way in the solution to Exercise 30(ii).

Exercise 25 (i) $2178 = 2 \times 1089$.

(ii) $21978 = 2 \times 10989$.

(iii) $21999978 = 2 \times 10999989$ and $21782178 = 2 \times 10891089$.

(iv) Elementary algebra shows that in base b we have

$$[1 \times b^3 + 0 \times b^2 + (b - 2) \times b + (b - 1)] \times (b - 1)$$
$$= (b - 1) \times b^3 + (b - 2) \times b^2 + 0 \times b + 1$$

and

$$[1 \times b^3 + 0 \times b^2 + (b - 2) \times b + (b - 1)] \times 2$$
$$= 2 \times b^3 + 1 \times b^2 + (b - 3) \times b + (b - 2).$$

So $\quad 1\ 0\ (b - 2)\ (b - 1)_{\text{base } b}$ is a $(b - 1)$-flip (base b), and

$$1\ 0\ (b - 2)\ (b - 1)_{\text{base } b} \times 2 = 2\ 0\ (b - 3)\ (b - 2)_{\text{base } b}.$$

Also

$$[2 \times b^3 + 0 \times b^2 + (b - 3) \times b + (b - 2)] \times \tfrac{1}{2}(b - 2)$$
$$= (b - 2) \times b^3 + (b - 3) \times b^2 + 1 \times b + 2,$$

so if b is an even number ≥ 4, then $2\ 0\ (b-3)\ (b-2)_{\text{base } b}$ is a $\frac{1}{2}(b-2)$-flip (base b).

Exercise 26 (ii) Suppose $3 \times (a \ldots z) = z \ldots a$. The left hand column shows that $a \leq 3$. Moreover if $a = 1$, then $3 \leq d \leq 5$; if $a = 2$, then $6 \leq d \leq 8$; and if $a = 3$, then $d = 9$. But the right-hand column shows that if $a = 1$ then $d = 7$, if $a = 2$ then $d = 4$, and if $a = 3$ then $d = 1$. So there are no 3-flips at all.

Exercise 27 (ii) Suppose $2 \times (a \ldots z) = z \ldots a$. The right-hand column shows that a is even. The left-hand column shows that $a \leq 4$, so $a = 2$ or $a = 4$. Moreover if $a = 2$, then $4 \leq d \leq 5$; and if $a = 4$, then $8 \leq d \leq 9$. But the right-hand column shows that if $a = 2$ then $d = 1$ or 6, and if $a = 4$ then $d = 2$ or 7. So there are no 2-flips at all.

8 Flips in other bases

*If digressions can bring us knowledge of new truths,
why should they trouble us? [. . .] how do we know that
we shall not discover curious things that are more
interesting than the answers we originally sought?*

(Galileo, *Two new sciences*)

Numbers are usually written in base 10. But you probably know already that they can also be written in other bases. One of the questions we raised at the end of Chapter 5 was the following.

Question D. *What happens if the numbers you work with are written in other bases? For example, if you write all numbers in base 9, are there any 8-flips at all? If so, can you find them all?*

We are so much more familiar with numbers written in base 10 that it was natural to start by looking at 'flips (base 10)'. But it is often instructive to investigate what happens in other bases.

However you may not have had much practice at doing arithmetic in other bases. The underlying idea is very simple. For example,

$$\text{two hundred and six} = 2 \times 10^2 + 0 \times 10 + 6,$$

so we write for short

$$\text{two hundred and six} = 2\,0\,6_{\text{base 10}}.$$

But it is just as true to say that

$$\text{two hundred and six} = 2 \times 9^2 + 4 \times 9 + 8$$

$$= 2\,4\,8_{\text{base 9}}.$$

When doing arithmetic in base 9 you add and subtract as in base 10. But you have to remember that all your numbers are written in base 9 so that you have to 'carry' *multiples of 9* (and not multiples of 10 as you are used to doing when your numbers are written in base 10). Thus $8 + 3 = 12_{\text{base 9}}$ and $37_{\text{base 9}} + 63_{\text{base 9}} = 101_{\text{base 9}}$.

It is worth making the effort to get what you can out of this short chapter even if you are not used to working in other bases. But the final two chapters of this investigation do not depend on the present chapter in any way. So if you get stuck, you can start afresh at the beginning of Chapter 9.

Exercises 28 (i) Find an 8-flip (base 9) with four digits.
(ii) Compare this 8-flip (base 9) with the unique four-digit 9-flip (base 10).
(iii) Guess a likely 7-flip (base 8) with four digits. Check that your guess really is a 7-flip (base 8).
(iv) Will the same idea work in every base? Would it work in base 61? What about base 734? What about base b?

Exercise 29 (i) In Exercise 28(i) you found *one* 8-flip (base 9) with four digits. Are there any other 8-flips (base 9) with four digits, or is yours the only one?
(ii) Can you guess any 8-flips (base 9) with five digits? Check that your guess really is an 8-flip (base 9). Are there any other 8-flips (base 9) with five digits, or is yours the only one?

Exercise 30 (i) You should now be in a position to fill in the table in Fig. 8.1.

Number of digits (base 9)	Complete list of all 8-flips (base 9)
1	
2	
3	
4	
5	

Fig. 8.1

(ii) What do you expect the next few entries in this table to be? Fill in your guesses in the table in Fig. 8.2; then check to see whether your guesses were correct.
(iii) Give a simple recipe which will generate all 8-flips (base 9).
(iv) Find a convincing mathematical explanation of why your recipe produces *all* 8-flips (base 9) without missing any out.

Exercises 31 In base 10 you found just one 4-flip with four digits, namely 2178.
(i) Does this example of a 4-flip (base 10) with four digits suggest

Number of digits (base 9)	Your *guessed* list of all possible 8-flips (base 9)
6	
7	
8	
9	
10	

Fig. 8.2

any likely candidates for a ?-flip (base 9) with four digits? Check your guess carefully.

(ii) Does the unique 4-flip (base 10) with four digits suggest any likely candiates for a ?-flip (base 8) with four digits? Check your guess very carefully.

(iii) How about four-digit flips in base 7, base 6, and so on?

The next problem is slightly different from previous exercises in that you are not told exactly what to do. But you have now accumulated so much experience that it should not be too difficult to plan your own method of attack.

Project Let $b \geq 2$ be any whole number. Find all flips in base b.

Hints to exercises

28. (i) Imitate Exercise 6, but this time work in base **9**, and multiply by **8**.
   (iv) Work in base b and use elementary algebra.
30. (iv) Look back at your solutions to Exercise 15 and Exercise 24.
31. (i) **4**-flip (base **10**) $\leftrightarrow$ **?**-flip (base **9**).
   (ii) **4**-flip (base **10**) $\leftrightarrow$ **?**-flip (base **8**).

Project Exercise 31 should suggest that there is a slight difference between what happens when b is even (base 8, base 10, base 12, and so on), and what happens when b is odd (base 9, base 7, base 11, and so on).

Solutions

Exercise 28 (i) $1078_{\text{base }9}$.
   (iii) $1067_{\text{base }8}$.
   (iv) In the solution to Exercise 25(iv) we saw that $(1 \times b^3 + 0 \times b^2 + (b - 2) \times b + (b - 1)) \times (b - 1) = (b - 1) \times b^3 + (b - 2) \times b^2 + 0 \times b + 1$, so $1\ 0\ (b - 2)\ (b - 1)_{\text{base }b}$ is a $(b - 1)$-flip (base b).

Exercise 29 (i) Let $abcd_{\text{base } 9}$ be an 8-flip (base 9). As in Exercise 16, $8 \times d$ must end in a. But since we are working in base 9 we can no longer conclude that 'a must be even'. (For example, $8 \times 2 = 17_{\text{base } 9}$.) So we go back to the method used in Exercise 6. $(8 \times a + ?)_{\text{base } 9}$ must not spill over into the next column, so $a = 1$, $? = 0$, and $d = 8$. But then nothing is carried over from the second to the first column, so $b \leqslant 1$. If $b = 1$, then $8 \times b = c = 8$ since nothing can be carried over from the third column. But then the third column goes wrong, since $8 \times c + 7$ does not end in $b = 1$. Hence $b = 0$; and since $8 \times c + 7$ ends in b, c must be 7, so $abcd = 1078_{\text{base } 9}$.

(ii) There is precisely one, namely $10878_{\text{base } 9}$.

Exercise 30 (iii) The basic 8-flips (base 9) are 0, 1078, 10878, 108878, ... These may be strung together in any way, as long as the blocks are arranged symmetrically about the middle, as in

$$107800108887801087810780107810878010888780001078$$

The fact that these are the only possible 8-flips (base 9) emerges when one examines an *unknown* 8-flip with an *unspecified* number of digits.

$$
\begin{array}{l}
\quad 1\ 0 \qquad\qquad\qquad 7\ 8 \\
\text{~~a~~ ~~b~~}cdef \ldots uvwx\text{~~y~~}\,\text{~~z~~} \quad \times \\
\qquad\qquad\qquad\qquad\quad 8 \\
\hline
\quad 8\ 7 \qquad\qquad\qquad 0\ 1 \\
\text{~~z~~}\,\text{~~y~~}xwvu \ldots fedc\,\text{~~b~~}\ \text{~~a~~}
\end{array}
$$

As usual we get $a = 1$, $z = 8$, $b = 0$, $y = 7$, and either $c = 7$ or $c = 8$. If $c = 7$, then $x = 0$, $d = 8$, $w = 1$, and our unknown 8-flip looks like $1078ef \ldots uv1078$. But then the middle section $ef \ldots uv$ has fewer digits than the original 8-flip and satisfies $8 \times (ef \ldots uv) = vu \ldots fe$. If $c = 8$, then $x = 8$, and $d = 7$ or $d = 8$. If $d = 7$, then $w = 0$, $e = 8$, $v = 1$ and our unknown 8-flip looks like $10878f \ldots u10878$. Again the middle section $f \ldots u$ has fewer digits than the original 8-flip and satisfies $8 \times (f \ldots u) = u \ldots f$. If $d = 8$, then $w = 8$, and $e = 7$ or $e = 8$. Continuing in this way we see that our unknown 8-flip looks like $1088 \ldots 8 \ldots 8 \ldots 8878$. Either the string of 8s at the beginning (namely $cd \ldots$) continues until it links up with the string of 8s at the end (namely $\ldots wx$), or the string of 8s at the beginning comes to an end before the two strings join up. In the first case we get one of our basic 8-flips $108888 \ldots 88878$. In the second case the string of 8s at the beginning is followed by a 7 and then an 8, and the string of 8s at the end is preceded by a 0 and a 1; but then our unknown 8-flip looks like $1088 \ldots 878jk \ldots pq1088 \ldots 878$. Again the middle section $jk \ldots pq$ has fewer digits than the original 8-flip and satisfies $8 \times (jk \ldots pq) = qp \ldots kj$. We can then apply exactly the same reasoning to the middle section $ef \ldots uv$, or $f \ldots u$, or $jk \ldots pq$ *as long as we realize that there is nothing to stop the first digit of this middle section being zero.*

Exercise 31 In the solution to Exercise 25(iv) we saw that the 4-flip $2178_{\text{base } 10}$ is a special case of a general phenomenon: namely that $2\,1\,(b-3)\,(b-2)_{\text{base } b}$ is always a $\frac{1}{2}(b-2)$-flip (base b).

(i) So when $b = 9$, the 4-flip $2178_{\text{base } 10}$ corresponds to the '$3\frac{1}{2}$-flip' $2167_{\text{base } 9}$.

You might like to investigate 'x-flips, when x is not a whole number' for yourself, but they do not really count as flips in the sense in which we have been using the term.

(ii) When $b = 8$, the 4-flip $2178_{\text{base } 10}$ corresponds to the 3-flip $2156_{\text{base } 8}$.

(iii) When $b = 6$ we get a 2-flip $2134_{\text{base } 6}$, and when $b = 4$ we get a 1-flip $2112_{\text{base } 4}$.

9 Counting 9-flips with *n* digits

Half the secret of resistance to disease is
cleanliness, the other half is dirtiness.
(Anon.)

You should by now feel that you would probably be able to answer almost any question about flips in any base provided you could answer the corresponding question about 9-flips (base 10). In Exercise 15(iii) and Exercise 24(iv) you probably gave your 'recipes' for constructing all 9-flips (base 10) and all 4-flips (base 10) *in words*. This may have been good enough to enable you to write down all 9-flips (base 10) with (say) sixteen digits. But it is not good enough. For example, it does not provide an answer to the very first of the general questions we posed way back in Chapter 5.

Question A *Is it possible to say exactly how many 9-flips (base 10) there are with precisely n digits?*

The last three chapters of this extended investigation explore Question *A* and the mathematics which is needed to answer it in a satisfactory way.

Exercise 32 (i) Write down the complete list of all 9-flips (base 10) with ten digits.

(ii) Write down the complete list of all 9-flips (base 10) with eleven digits.

(iii) Write down the complete list of all 9-flips (base 10) with twelve digits.

(iv) How many 9-flips are there with twenty four digits?

(v) How many 9-flips are there with forty four digits?

You probably had a little difficulty with Exercise 32(iv), to say nothing of Exercise 32(v)! But if the 9-flips (base 10) can all be constructed according to the simple recipe you gave in Exercise 15, then it should surely be possible to find a simple way of calculating the 24th term, the 44th term, and in general the nth term of the sequence (see Fig. 9.1). This is what we shall be trying to do in this chapter and

Number of digits	1	2	3	4	5	6	7	8	9	10	11	12	13	14	15	16	·
Number of 9-flips	1?	0	0	1	1	1	1	2	2	3	3	5	?	?	?	?	·

Fig. 9.1

the next. To save time and space we need a short way of referring to

'the number of 9-flips with *n* digits'.

So we shall write this number as f_n (*f* for flips, and *n* for the **n**umber of digits).

Exercise 33* Fill in the table in Fig. 9.2 as far as $n = 14$.

n	Complete list of 9-flips with *n* digits	f_n
1	0	1
2		
3		
4		
5		
6		
7		
8		
9		
10		
11		
12		
13		
14		

Fig. 9.2

Filling in a table like this is only useful if it suggests some more efficient way of calculating f_n for larger values of *n*. The method you actually used in Exercise 33 is much too much like hard work. First you had to write out the complete list of all 9-flips with *n* digits: only then could you begin to count the number of 9-flips with *n* digits. This method is very wasteful, since you only want to know *how many* 9-flips there are with *n* digits—you don't particularly want a complete list of them all. Worse still, as the list gets longer it is much too easy to make a silly mistake without realizing it (either missing some out, or counting some more than once). For example, there are *eighty nine*

9-flips with twenty four digits, *nine hundred and eighty eight* 9-flips with thirty four digits, and *ten thousand nine hundred and forty seven* 9-flips with forty four digits. Unless you find some better way of counting them all you are bound to miss some. And what is worse, you would have no way of knowing that you had missed them!

It is easy to explain in words what the value of f_n is: namely 'f_n is equal to the number of 9-flips with precisely *n* digits'. This description specifies an *endless* sequence of numbers f_n, one for each value of *n*

$$f_1, f_2, f_3, f_4, f_5, f_6, f_7, f_8, f_9, f_{10}, f_{11}, f_{12}, f_{13}, \ldots.$$

You have already worked out the first fourteen of these numbers, or the first fourteen *terms* as we shall call them. But you had to work them all out *the hard way*, by first writing out a complete list of 9-flips with *n* digits. What you want is a simple recipe, or formula, so that you can work out any one of the terms f_n which you may happen to need *by doing as little work as possible*.

How can one find a more efficient way of calculating f_n for larger values of *n*? One approach which is often very effective is to look for a formula, or equation, which shows how *the nth term f_n* of the sequence

$$f_1, f_2, f_3, f_4, f_5, \ldots, f_n, \ldots$$

can be calculated as soon as one knows the terms just before it. In other words, to calculate f_{15} *efficiently* you could try to find an equation

$$f_{15} = \text{'some expression involving } f_{14}, f_{13}, f_{12}, \text{ and so on'}.$$

To calculate the nth term f_n efficiently you need an equation

$$f_n = \text{'some expression involving } f_{n-1}, f_{n-2}, f_{n-3}, \text{ and so on'}.$$

An equation of this kind is called a **recurrence relation for the sequence $f_1, f_2, f_3, \ldots.$** As soon as you discover a recurrence relation for your sequence $f_1, f_2, f_3, \ldots$ you will immediately be able to *calculate* f_{15} without first writing down a complete list of all 9-flips with 15 digits: just substitute the values of $f_{14}, f_{13}, f_{12}, \ldots$ (which you already know) in the right-hand side of the recurrence relation

$$f_{15} = \text{'some expression involving } f_{14}, f_{13}, f_{12}, \text{ and so on'}.$$

You will then know the value of f_{15}, and so will be able to calculate f_{16} by substituting the values of $f_{15}, f_{14}, f_{13}, \ldots$ (which you now know) in the right-hand side of the recurrence relation

$$f_{16} = \text{'some expression involving } f_{15}, f_{14}, f_{13}, \text{ and so on'}.$$

You can then repeat this procedure over and over again to calculate f_{17}, f_{24}, f_{34}, f_{44}, or f_n for any particular value of n.

The next exercise is not directly relevant to Question *A*, but if you have not met recurrence relations before it should give you some practice in a relatively familiar context.

Exercise 34 Just suppose you did not know, or had forgotten, the formula for the sum of the angles in a polygon with *n* vertices. How could you work out such a formula? First of all you need a short way of referring to 'the sum of the angles in a polygon with *n* vertices'. It seems reasonable to write this number as a_n (*a* for angle-sum, *n* for the **n**umber of vertices).

(i) What happens when $n = 3$? In this case 'a polygon with *n* vertices' is just a triangle, so a_3 is the angle-sum in a triangle. Hence $a_3 = 180°$. And what happens when $n = 4$? In this case 'a polygon with *n* vertices' is just a quadrilateral, and you probably know that the sum of the angles in a quadrilateral is 360°, so $a_4 = 360°$. But just supposing you didn't know this already. *How could you work it out from scratch?* One way is to cut the quadrilateral into two triangles Fig. 9.3(i). It is then obvious that the sum of the angles in the quadrilateral is equal to the sum of the angles in the two triangles. So

$$a_4 = a_3 + a_3 = a_3 + 180.$$

(ii) What happens when $n \geqslant 5$. We can then cut any polygon with *n* vertices into two pieces, one a triangle and the other a polygon with $n - 1$ vertices (Fig. 9.3(ii)). It is then obvious that

$a_n = $ (sum of angles in polygon with *n* vertices)

$ = $ (sum of angles in polygon with $n - 1$ vertices)

$ + $ (sum of angles in triangle).

$\therefore a_n = a_{n-1} + 180.$

(iii) We have just found a *recurrence relation* for the sequence

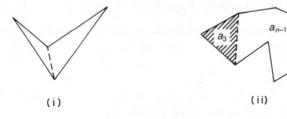

(i)   (ii)

Fig. 9.3

$a_3, a_4, a_5, \ldots.$ Use this recurrence relation to complete the table in Fig. 9.4.

Number of vertices of polygon = n	3	4	5	6	7	8	9	10	11	12	13
Sum of angles in polygon = a_n	180										

Fig. 9.4

(iv) In Part (ii) we found the basic recurrence relation $a_n = a_{n-1} + 180$. But if you wanted to calculate a_{1089} (that is, the angle-sum in a polygon with 1089 vertices), then this recurrence relation would be very inefficient. For in order to find a_{1089} you would first need to know a_{1088}; and to find a_{1088} you would first need to know a_{1087}; and so on. It would be much easier if you had a *formula* for a_n which expressed a_n in terms of *n* alone. Find such a formula.

The remaining exercises in this and the next two chapters give you the chance to use recurrence relations to answer Question *A* completely by finding the value of f_n **for every *n*.**

In Exercise 15 you should have noticed that almost all 9-flips are obtained by sticking shorter 9-flips together. In fact *for each $n \geq 4$, there is just one 9-flip which is **not** obtained by sticking shorter 9-flips together, namely*

$$\underbrace{1099\ldots9989.}_{all\ 9s}$$

All other 9-flips look like

$1089\ldots1089,$ or $10989\ldots10989,$ or $109989\ldots109989,$ or $\ldots.$

But wait a minute! Isn't this just the kind of observation which might help you find an equation of the form

$f_n =$ some expression involving earlier terms $f_{n-1}, f_{n-2}, f_{n-3}$, etc.?

We shall now explore this idea by doing some detailed calculations for particular values of *n*. Let's start by looking at 9-flips with $n = 10$ digits (Fig. 9.5).

Exercise 35* Now it is you turn to do the same for 9-flips with $n = 14$ digits. Complete the table in Fig. 9.6.

These two examples should suggest a way of counting all 9-flips

9-flips with ten digits	9-flips	Number
(1) There is just one such 9-flip which is not obtained by sticking shorter 9-flips together, namely 1099999989.	1099999989	1
(2) There is just one such 9-flip with 1089 at the beginning and end, namely 1089001089.	1089001089	1
(3) There is just one such 9-flip with 10989 at the beginning and end, namely 1098910989.	1098910989	1
		$f_{10} = 3$

Fig. 9.5

9-flips with fourteen digits	9-flips	Number
(1) There is *just one* such 9-flip which is not obtained by sticking shorter 9-flips together, namely	_____	1
(2) There are *just three* such 9-flips with 1089 at the beginning and end, namely	1089 ____ 1089 1089 ____ 1089 1089 ____ 1089	3
(3) There are _____ such 9-flips with 10989 at the beginning and end, namely		____
(4) There are _____ such 9-flips with 109989 at the beginning and end, namely		____
(5) There is _____ such 9-flip with 1099989 at the beginning and end, namely		____
		$f_{14} =$ ____

Fig. 9.6

with *n* digits. In Exercise 35 you worked out f_{14} by using the equation

f_{14} = (number 9-flips not constructed from shorter 9-flips)

+ (number of 9-flips which begin and end with 1089)

+ (number of 9-flips which begin and end with 10989)

+ (number of 9-flips which begin and end with 109989)

+ (number of 9-flips which begin and end with 1099989).

And we could obviously calculate f_n for other values of *n* by making use of the same idea, namely

f_n = (number of 9-flips not constructed from shorter 9-flips)

+ (number of 9-flips which begin and end with 1089)

+ (number of 9-flips which begin and end with 10989)

+ (number of 9-flips which begin and end with 109989)

+ . . . .

But this formula will only be useful if we can find a simple way of working out the values of the terms on the right-hand side.

The first term is easy enough. For each value of *n* there is *just one* 9-flip which is not constructed by sticking shorter 9-flips together. But how on earth are we going to calculate the number of 9-flips with *n* digits which begin and end with 1089, or with 10989, or . . . ?

Exercise 36* (There are no obvious answers to this exercise. Give serious thought to the questions it raises, try to answer them, and then read on.)

(i) Suppose $n \geqslant 8$. Can you find some way of counting the number of 9-flips with *n* digits which begin and end with 1089: that is, 9-flips which look like this

$$1089 \underbrace{\hspace{4cm}}_{n-8 \text{ digits}} 1089?$$

(ii) Suppose $n \geqslant 10$. Can you find some way of counting the number of 9-flips with *n* digits which begin and end with 10989: that is 9-flips which look like this

$$10989 \underbrace{\hspace{4cm}}_{n-10 \text{ digits}} 10989?$$

To count the number of 9-flips with *n* digits which begin and end with 1089, that is, 9-flips which look like

$$1089abc \ldots z1089,$$

you only need to count the number of ways in which the middle $n - 8$ digits $abc \ldots z$ can be filled in. But

$$
\begin{array}{c|c|c}
1089 & abc \ldots z & 1089 \\
\hline
9801 & z \ldots cba & 9801
\end{array} \times 9
$$

so the number $abc \ldots z$ formed by the middle $n - 8$ digits must itself satisfy

$$
\begin{array}{c}
abc \ldots z \\
\hline
z \ldots cba
\end{array} \times 9
$$

So it looks as though the number formed by the middle $n - 8$ digits has to be *a genuine 9-flip with $n - 8$ digits!* If this were true then the number of 9-flips with n digits which begin and end with 1089 would be exactly equal to the number of 9-flips with $n - 8$ digits, namely f_{n-8}.

Exercise 37* Unfortunately this is not quite correct. Can you find the mistake?

Let's go over the whole argument again. To count the number of 9-flips which have n digits and which look like

$$1089abc \ldots z1089$$

it is certainly enough to count the number of ways in which the middle $n - 8$ digits can be filled in. And the number $abc \ldots z$ formed by the middle $n - 8$ digits certainly satisfies

$$\frac{\begin{array}{r} abc \ldots z \\ 9 \end{array} \times}{z \ldots cba}$$

So it looks as if $abc \ldots z$ must be a 9-flip. But wait! We have assumed right from the start that *the only 'proper' 9-flip which begins with a zero is the 9-flip '0' with just one digit.* But there is nothing to prevent the first digit a of the middle bit $abc \ldots z$ from being a zero. For example, when we made a list of all 9-flips with ten digits we found exactly one 9-flip which begins and ends with 1089, namely

$$1089\mathbf{00}1089,$$

and the middle $n - 8$ digits here just give the number **00**. So to count the number of ways the middle $n - 8$ digits can be filled in you have to count 'improper' 9-flips $abc \ldots z$ with $n - 8$ digits, *which may begin with 0,* and not just 'proper' 9-flips $abc \ldots z$ (which could begin with zero only if $n - 8 = 1$). When we say that '$abc \ldots z$ is an *improper* 9-flip' it will simply mean that multiplying by 9 produces the same digits in reverse order:

$$\frac{\begin{array}{r} abc \ldots z \\ 9 \end{array} \times}{z \ldots cba}$$

In particular, if $a = 0$ then $z = 0$ (why); if $a = b = 0$, then $y = z = 0$ (why?); and so on. Notice that though an improper 9-flip *may* begin with zero, *it doesn't have to:* 10989, 0109890, 108901089, and 0010890108900 are all improper 9-flips. So **every proper 9-flip counts as an improper 9-flip too.**

If you are going to look for some easy way of counting 'the number of improper 9-flips with $n - 8$ digits' you will probably need a short way of referring to this number. Let i_n stand for the 'the number of improper 9-flips with n digits' (*i* for **i**mproper, and *n* for the **n**umber of digits).

It is worth drawing up a table of 'improper 9-flips with n digits' for small values of n just to make sure that you have got hold of the basic idea.

Exercise 38   Extend the table in Fig. 9.7 as far as $n = 10$.

n	i_n	Complete list of improper 9-flips with n digits
1	1	0
2	1	00
3	1	000
4	2	0000, 1089
5	2	00000, 10989
6	3	000000, 010890, 109989
7	.	.
8	.	.
9	.	.
10	.	.

Fig. 9.7

We can now go back to our formula for the number of (proper) 9-flips with n digits:

f_n = (number of proper 9-flips not constructed from shorter 9-flips)

   + (number of proper 9-flips which begin and end with 1089)

   + (number of proper 9-flips which begin and end with 10989)

   + (number of proper 9-flips which begin and end with 109989)

   + ...

Exercise 39*   (i) The above equation for f_n is written in words.

Rewrite it as an equation involving only the symbols f_n, i_{n-8}, i_{n-10}, i_{n-12}, ...

(ii) Write down the corresponding equation for f_{n-2}.

(iii) Hence obtain an equation involving just the three numbers f_n, f_{n-2}, i_{n-8}.

Exercise 40 The equation you obtained in Exercise 39(iii) says that, as soon as you know f_{n-2} and i_{n-8}, you can use these to calculate f_n. Use this equation and the relevant values in the table in Fig. 9.8 to fill in the value of f_9. Then fill in the values of i_2, i_3, . . . , i_{10} which you found in Exercise 38 and use these to calculate and fill in the values of f_n as far as $n = 18$.

n	1	2	3	4	5	6	7	8	9	10	11	12
i_n	1											
f_n	0	0	0	1	1	1	1	2				

Fig. 9.8

You are looking for a recurrence relation of the general form

$$f_n = \text{'some expression involving } f_{n-1}, f_{n-2}, f_{n-3}, \ldots \text{'}.$$

But as yet the nearest you have got to such a recurrence relation is the equation

$$f_n = f_{n-2} + i_{n-8}.$$

Unfortunately this equation involves the number i_{n-8}, which is not one of the terms of the sequence $f_1, f_2, f_3, \ldots$. But the equation could still be quite useful if only you could find a convenient way of calculating i_{n-8} for larger values of n.

Exercise 41* (i) The method you used to find the equation $f_n = f_{n-2} + i_{n-8}$ can also be used to find an equation for i_n. Go back and try to imitate the method you used in Exercise 39 for f_n to find a similar equation for i_n.

(ii) Compare your own solution to Part (i) with the approach outlined in the hint.

Exercise 42* You now have two equations: one expressing f_n as the sum of f_{n-2} and i_{n-8}, and the other expressing i_n as the sum of f_n and i_{n-2}. Use these two equations to extend the table in Fig. 9.9 as far as $n = 25$.

n	1	2	3	4	5	6	7	8	9	10	11	12	13	. . .
i_n	1	1	1	2	.	.	.							
f_n	0	0	0	1	1	1	1	2	2	.	.	.		

Fig. 9.9

Exercise 43   At the beginning of this chapter I claimed that

$$f_{24} = 89, \qquad f_{34} = 988, \qquad f_{44} = 10947.$$

But at least one of these values is wrong. Find the correct values.

This method of calculating f_n is clearly a great improvement on your original method, in which you first worked out all the 9-flips with n digits and then simply counted how many 9-flips you had managed to find. But this improved method is still only a partial answer to Question A. If you did Exercise 43 by hand then you may have noticed already that your improved method has its own shortcomings. We shall look at a specific example before discussing these shortcomings.

Suppose you wanted to calculate f_{100} or f_{1000}. The first of these would certainly be possible by hand, but it would take quite a long time and you would almost certainly make a silly arithmetical mistake somewhere along the line. You could avoid making such arithmetical mistakes by using a calculator, but it would still take a long time to find f_{100}—unless, of course, you used a programmable calculator.

Exercise 44   Get hold of a programmable calculator and find out how to write a simple program to calculate f_n. Then find f_{100}.

The idea of using recurrence relations to calculate f_n has paid off in that you now have a much improved way of calculating f_n. But your method has two slightly undesirable features. Consider the above example of trying to find f_{100}.

(1) If you do the necessary calculations by hand, you must first work out *ninety five* numbers which you don't really need, namely

$$f_2, f_4, f_6, f_8, \ldots, f_{96}, f_{98}$$

and

$$i_2, i_4, i_6, i_8, \ldots, i_{90}, i_{92}.$$

All of this takes a very long time and makes it more likely that you will make some silly arithmetical mistake.

(2) If you save time and avoid making such silly arithmetical mistakes by using a programmable calculator, then the calculator still has to work out these *ninety-five* unwanted numbers first; and even then the numbers soon get too large to be given exactly on a calculator.

The first of these undesirable features would become slightly less unpleasant if you could somehow find a *genuine* recurrence relation for f_n—that is, an equation for f_n which does not involve i_{n-8} or any of the *i*s, but only the *f*s.

Exercise 45* (i) You found the equation $f_n = f_{n-2} + i_{n-8}$ in a slightly roundabout way, by first showing that

$$f_n = 1 + i_{n-8} + i_{n-10} + i_{n-12} + \dots$$

and then subtracting the corresponding equation for f_{n-2},

$$f_{n-2} = 1 + i_{n-10} + i_{n-12} + i_{n-14} + \dots.$$

Look carefully at the equation $f_n = f_{n-2} + i_{n-8}$ and try to find a quick way of proving it *directly* with no intermediate steps at all.

(ii) Now do the same for the formula $i_n = f_n + i_{n-2}$.

(iii) Use these two equations to derive a genuine recurrence relation for the sequence $f_1, f_2, f_3, \dots$ involving just the four terms f_n, $f_{n-2}, f_{n-4}, f_{n-8}$.

The recurrence relation you found in Exercise 45(iii) shows that you no longer need to worry about improper 9-flips. If you had been very clever, or very lucky, then it is just possible that you might have noticed this ages ago. But I doubt it! In fact it is not at all clear that you would ever have managed to *prove* that f_n satisfies this very simple recurrence relation if you had not allowed yourself to introduce improper 9-flips first as a kind of stepping-stone.

*[You may by now have guessed that the fs actually satisfy an even simpler recurrence relation than the one you found in Exercise 45(iii). But even if you have made such a **guess**, you probably cannot yet **prove** that this simpler recurrence relation works for every n. So for the moment we shall stick to the recurrence relation which you found in Exercise 45(iii) and which you have proved holds for every n. But if you do suspect that the fs satisfy a simpler recurrence relation you*

should keep an eye out for further evidence that might help you check whether your guess is correct.]

Hints to exercises

32. (i) There are three in all.
    (ii) There are three in all.
    (iii) There are five in all.
34. (iv) $a_5 = a_4 + 180 = (a_3 + 180) + 180 = a_3 + 2 \times 180 = 3 \times 180$.
    $a_6 = a_5 + 180 = \ldots$.
36. (i) In how many ways can the middle $n - 8$ digits be filled in?
    (ii) In how many ways can the middle $n - 10$ digits be filled in?
37. Suppose

$$\begin{array}{r} 1089abc\ldots z1089 \\ \times \quad 9 \\ \hline 9801z\ldots cba9801 \end{array}$$

What can you say about a?

39. (i) Number of proper 9-flips not constructed from shorter 9-flips = ____ .
    Number of proper 9-flips which begin and end in 1089 = ____ .
    Number of proper 9-flips which begin and end in 10989 = ____ .
    And so on. Hence $f_n =$ _____ .

41. (i) i_n = (number of improper 9-flips with n digits)
    = (number of improper 9-flips with no zeros at the beginning)
    + (number of improper 9-flips with one zero at the beginning)
    + (number of improper 9-flips with two zeros at the beginning)
    + ...

    (*a*) First find the value of each bracket on the right-hand side of this equation, and so obtain an equation for i_n.

    (*b*) Then write down the corresponding equation for i_{n-2}, and so obtain an equation involving only i_n, i_{n-2}, and f_n.

44. If you want to find f_{100} using the equation $f_{100} = f_{98} + i_{92}$, then you first have to work out f_{98} and i_{92}. And to find f_{98} you first need to know f_{96} and i_{90}. And so it goes on. Thus you must somehow get your calculator to

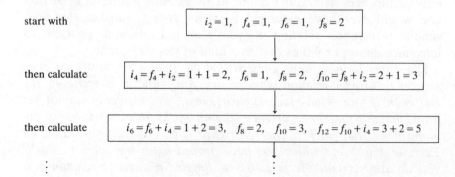

start with | $i_2 = 1$, $f_4 = 1$, $f_6 = 1$, $f_8 = 2$

then calculate | $i_4 = f_4 + i_2 = 1 + 1 = 2$, $f_6 = 1$, $f_8 = 2$, $f_{10} = f_8 + i_2 = 2 + 1 = 3$

then calculate | $i_6 = f_6 + i_4 = 1 + 2 = 3$, $f_8 = 2$, $f_{10} = 3$, $f_{12} = f_{10} + i_4 = 3 + 2 = 5$

45. (i) $f_n =$ (number of 9-flips with first and last block 1089) + (number of 9-flips whose first block has at least one 9 in the middle)

(iii) The first formula gives $f_{n-2} = f_{n-4} + i_{n-10}$. The second gives $i_{n-8} = f_{n-8} + i_{n-10}$.

Solutions

Exercise 32   (i) 1099999989, 1089001089, 1098910989.

(ii) 10999999989, 10890001089, 10989010989.

(iii) 109999999989,   108900001089,   108910891089,   109890010989, 109989109989.

(iv), (v) We shall come back to these later.

Exercise 33 $f_1 = 1$, $f_2 = 0$, $f_3 = 0$, $f_4 = 1$, $f_5 = 1$, $f_6 = 1$, $f_7 = 1$, $f_8 = 2$, $f_9 = 2$, $f_{10} = 3$, $f_{11} = 3$, $f_{12} = 5$, $f_{13} = 5$, $f_{14} = 8$.

Exercise 34   (iii)

n	3	4	5	6	7	8	9	10	11	12	13
a_n	180	360	540	720	900	1080	1260	1440	1620	1800	1980

(iv) $a_n = (n-2) \times 180$.

Exercise 35 $f_{14} = 8$.

Exercise 36   Suppose that $9 \times (1089abc \ldots z1089) = 9801z \ldots cba9801$. Nothing can be carried over from $9 \times a$ to the next column, so either $a = 0$ or $a = 1$. There is in fact no way of ruling out the possibility that $a = 0$, so we simply have to accept that the middle section $abc \ldots z$ may begin with one or more zeros. And though $abc \ldots z$ has $n - 8$ digits and satisfies $9 \times (abc \ldots z) = z \ldots cba$, if $a = 0$ then it is not a proper 9-flip.

Exercise 38 $i_7 = 3$, $i_8 = 5$, $i_9 = 5$, $i_{10} = 8$.

Exercise 39   (i) $f_n = 1 + i_{n-8} + i_{n-10} + i_{n-12} + \ldots$.

(ii) $f_{n-2} = 1 + i_{n-10} + i_{n-12} + \ldots$.

(iii) $f_n - f_{n-2} = i_{n-8}$.

Exercise 41   (i) Using the hint we get $i_n = f_n + f_{n-2} + f_{n-4} + \ldots$, so $i_{n-2} = f_{n-2} + f_{n-4} + \ldots$ and $i_n - i_{n-2} = f_n$.

Exercise 42

n	1	2	3	4	5	6	7	8	9	10	11	12	13	14	15	16	17	18	19	20	21	22	23	24	25
i_n	1	1	1	2	2	3	3	5	5	8	8	13	13	21	21	34	34	55	55	89	89	144	144	233	233
f_n	1	0	0	1	1	1	1	2	2	3	3	5	5	8	8	13	13	21	21	34	34	55	55	89	89

Exercise 43 $f_{24} = 89$, $f_{34} = 987$, $f_{44} = 10946$.

Exercise 44 $f_{100} = 7778742049$. (Your calculator will probably have started rounding the final digits by the time it gets to f_{100}.)

Exercise 45 (i) If a proper 9-flip with n digits begins and ends with 1089, then removing these two blocks gives rise to an improper 9-flip with $n - 8$ digits: so there are i_{n-8} 9-flips like this with n digits. If the first block of a 9-flip with n digits is not 1089, then it begins $109\ldots$, and the last block ends $\ldots 989$, so we can remove the third digit (9) and the third last digit (9) to get a proper 9-flip with $n - 2$ digits: so there are f_{n-2} 9-flips like this with n digits. Hence $f_n = i_{n-8} + f_{n-2}$.

(ii) If an improper 9-flip with n digits begins with a zero, then it must end with a zero, and we can remove these first and last zeros to get an improper 9-flip with $n - 2$ digits: so there are i_{n-2} improper 9-flips like this with n digits. If an improper 9-flip does not begin with a zero, then it is a proper 9-flip: so there are f_n improper 9-flips like this with n digits. Hence $i_n = i_{n-2} + f_n$.

(iii) $f_n = i_{n-8} + f_{n-2}$ and $f_{n-2} = i_{n-10} + f_{n-4}$ by (i), so $f_n - f_{n-2} = i_{n-8} + f_{n-2} - i_{n-10} - f_{n-4}$. But $i_{n-8} - i_{n-10} = f_{n-8}$ by (ii), so $f_n = 2f_{n-2} - f_{n-4} + f_{n-8}$.

10 Recurrence relations

I have yet to see any problem, however complicated,
which, when looked at in the right way, did not
become still more complicated.

(Poul Anderson in *New Scientist*, 1969)

It would of course be very nice if you could prove that the sequence $f_1, f_2, f_3, \ldots$ satisfies some surprisingly simple recurrence relation. But this would still not provide us with a very efficient way of calculating f_{1000} or $f_{1\,000\,000}$. What we need is not just another recurrence relation, however simple it may be, but rather

a formula for f_n in terms of n alone.

If we had such a formula, then we could calculate f_{1000} very quickly by simply substituting $n = 1000$ in the formula.

We met exactly the same idea when we were trying to find a_n, the angle-sum in a polygon with n vertices (Exercise 34). We first obtained the recurrence relation

$$a_n = a_{n-1} + 180.$$

And although it would not be difficult to calculate a_{1000} using this recurrence relation, one would first have to work out

$$a_4, a_5, a_6, a_7, \ldots, a_{998}, a_{999}$$

before finally getting a_{1000}. In practice no-one would dream of calculating a_{1000} in this way, because it is extremely easy to replace the recurrence relation by *a formula for a_n in terms of n alone*, namely

$$a_n = (n - 2) \times 180.$$

One can then simply substitute $n = 1000$ in this formula to discover that

$$a_{1000} = 998 \times 180.$$

This very simple example underlines the important fact that 'a formula for a_n in terms of n alone' is much more useful than a mere recurrence relation. But the simplicity of the example is a bit misleading. The simplest recurrence relation which you have so far managed to prove for the number of 9-flips with n digits is

$$f_n = 2f_{n-2} - f_{n-4} + f_{n-8},$$

and this is considerably more complicated than the recurrence relation $a_n = a_{n-1} + 180$ for the sum of the angles in a polygon with n vertices. So though it was fairly easy to translate the recurrence relation for a_n into the formula $a_n = (n-2) \times 180$, you should not really expect it to be quite as easy to translate the recurrence relation for f_n into a formula for f_n in terms of n alone.

To get some idea of the kind of formula you should be looking for and the way you can go about finding it we shall begin by looking at a sequence which is more complicated than the angle-sum sequence $a_3, a_4, a_5, \ldots$ but which is slightly less complicated than the flips sequence $f_1, f_2, f_3, \ldots$. The sequence we shall study is the famous *Fibonacci sequence*

$$0, 1, 1, 2, 3, 5, 8, 13, 21, 34, 55, \ldots.$$

Exercise 46 Guess the next five terms of this sequence.

We shall need a short way of referring to 'the nth term of the Fibonacci sequence'. Let F_n be the nth term of this sequence (F for Fibonacci, and n for the nth term). Then

$$F_1 = 0, \ F_2 = 1, \ F_3 = 1, \ F_4 = 2, \ F_5 = 3, \ F_6 = 5, \ F_7 = 8, \ldots$$

and so on. To answer Exercise 46 correctly you would have had to 'see' (or know already) that in the Fibonacci sequence
    each term is equal to the sum of the two preceding terms.
In the language of recurrence relations this simply says that the terms of the Fibonacci sequence satisfy the recurrence relation

$$F_n = F_{n-1} + F_{n-2}. \tag{10.1}$$

Exercise 47 There are in fact infinitely many different sequences all of which satisfy the recurrence relation (10.1).

(i) The following six sequences all satisfy (10.1). Fill in the next three terms in each sequence.

(a) $0, 0, 0, 0, 0, 0,$ ____, ____, ____, ...
(b) $0, 1, 1, 2, 3, 5,$ ____, ____, ____, ...
(c) $1, 0, 1, 1, 2, 3,$ ____, ____, ____, ...
(d) $1, 3, 4, 7, 11, 18,$ ____, ____, ____, ...
(e) $-1, -1, -2, -3, -5, -8,$ ____, ____, ____, ...
(f) $0, 2, 2, 4, 6, 10,$ ____, ____, ____, ...

(ii) Let $a = (1 + \sqrt{5})/2$ and $b = (1 - \sqrt{5})/2$. Show that the two sequences (g) and (h) below both satisfy the recurrence relation (10.1).

(g) 1, a, a^2, a^3, a^4, a^5, ...
(h) 1, b, b^2, b^3, b^4, b^5, ...

Any sequence which satisfies the recurrence relation (10.1) is completely determined as soon as we specify its first two terms: the third term is then equal to the sum of these first two terms, the fourth term is equal to the sum of the second and third terms, and so on. So the Fibonacci sequence is completely determined by the following two facts:

(1) $F_1 = 0$ and $F_2 = 1$,
(2) the nth term F_n is given by the recurrence relation (10.1).

The special sequences

$$1, a, a^2, a^3, a^4, a^5, \ldots \quad \text{and} \quad 1, b, b^2, b^4, b^5, \ldots$$

are particularly important for us not merely because they satisfy the recurrence relation (10.1), but because

the nth term of each of these sequences is given by an arithmetical expression involving n alone,

namely

$$n^{\text{th}} \text{ term} = a^{n-1} = \left(\frac{1 + \sqrt{5}}{2}\right)^{n-1} \quad \text{and} \quad n^{\text{th}} \text{ term} = b^{n-1} = \left(\frac{1 - \sqrt{5}}{2}\right)^{n-1}.$$

But what is the particular significance of these two numbers a and b? Couldn't one say exactly the same for *any* sequence 1, x, x^2, x^3, x^4, x^5, ... which happens to satisfy the recurrence relation (10.1)?

Exercise 48 Suppose 1, x, x^2, x^3, x^4, x^5, ... is a sequence which satisfies the recurrence relation (10.1). Find all possible values for the number x.

We shall now show how to use these two special sequences 1, a, a^2, a^3, a^4, a^5, ... and 1, b, b^2, b^3, b^4, b^5, ... to obtain a formula which expresses the nth term of the Fibonacci sequence in terms of n alone. The key idea is contained in the next exercise.

Exercise 49 Each of the sequences in Exercise 47 satisfies the recurrence relation (10.1).

(i) Notice that you can get each term of sequence (f) in Exercise 47 by multiplying the corresponding term of the Fibonacci sequence (b) by 2. Suppose you know that u_1, u_2, u_3, ... is a sequence which satisfies the recurrence relation (10.1). Show that the sequence cu_1, cu_2, cu_3, ... which you get by multiplying each term of the first sequence by the fixed number c, also satisfies the recurrence relation (10.1).

(ii) Notice that you can also get each term of the sequence (f) in Exercise 47 by adding the corresponding terms of the sequences (d) and (e). Suppose you know that the two sequences u_1, u_2, u_3, ... and v_1, v_2, v_3, ... both satisfy the recurrence relation (10.1). Show that the sequence $u_1 + v_1$, $u_2 + v_2$, $u_3 + v_3$, ... also satisfies the recurrence relation (10.1).

The two sequences 1, a, a^2, a^3, ... and 1, b, b^2, b^3, ... both satisfy the recurrence relation (10.1). So Exercise 49(i) says that if you multiply each term of the first sequence by some fixed number x, and each term of the second sequence by some fixed number y, then you will get two new sequences

$$x, \ xa, \ xa^2, \ xa^3, \dots \quad \text{and} \quad y, \ yb, \ yb^2, \ yb^3, \dots$$

which also satisfy the recurrence relation (10.1). Exercise 49(ii) then says that if you now add the corresponding terms in these two new sequences, then you will get another sequence

$$x + y, \ xa + yb, \ xa^2 + yb^2, \ xa^3 + yb^3, \dots$$

which satisfies the recurrence relation (10.1).

By varying the values you choose for the fixed numbers x and y you can in fact get infinitely many different sequences in this way, and each of the sequences you get will satisfy the recurrence relation (10.1). **Is it perhaps possible to choose the numbers x and y so that the sequence**
$$x + y, \ xa + yb, \ xa^2 + yb^2, \ xa^3 + yb^3, \dots$$

is identical to the Fibonacci sequence F_1, F_2, F_3, F_4, ...? If this were possible, then you would have discovered a formula for the nth term of the Fibonacci sequence in terms of n alone, namely

$$F_n = xa^{n-1} + yb^{n-1} = x\left(\frac{1 + \sqrt{5}}{2}\right)^{n-1} + y\left(\frac{1 - \sqrt{5}}{2}\right)^{n-1}.$$

Exercise 50* Recall that the Fibonacci sequence is completely determined by the following two properties:
(1) $F_1 = 0$ and $F_2 = 1$, and
(2) the sequence satisfies the recurrence relation (10.1).

The sequence $x + y$, $xa + yb$, $xa^2 + yb^2, \ldots$ satisfies the recurrence relation (10.1) no matter what values we choose for x and y. So this sequence will be identical to the Fibonacci sequence precisely when we choose x and y to satisfy the equations of (1): that is, such that $x + y = F_1 = 0$, and $xa + yb = F_2 = 1$.

(i) Use the known values of a and b to find values of x and y which satisfy these two equations.

(ii) Hence write down a formula which expresses the nth term of the Fibonacci sequence in terms of n alone.

We now return to the original problem of finding a formula which expresses f_n (the number of 9-flips with precisely n digits) in terms of n alone. We shall do this by imitating, step for step, the method which you have just used to get a formula which expresses the nth Fibonacci number F_n in terms of n alone.

Our old recurrence relation

$$f_n = 2f_{n-2} - f_{n-4} + f_{n-8}$$

for the sequence $f_1, f_2, f_3, \ldots$ shows that each *even*-numbered term f_{2n} satisfies the equation

$$f_{2n} = 2f_{2(n-1)} - f_{2(n-2)} + f_{2(n-4)}.$$

So f_{2n} depends only on the even-numbered terms $f_{2(n-1)}, f_{2(n-2)}, f_{2(n-4)}$ which come before it, and not on the odd-numbered terms. Similarly each *odd*-numbered term f_{2n-1} satisfies the equation

$$f_{2n-1} = 2f_{2(n-1)-1} - f_{2(n-2)-1} + f_{2(n-4)-1}.$$

So f_{2n-1} depends only on the odd-numbered terms $f_{2(n-1)-1}, f_{2(n-2)-1}, f_{2(n-4)-1}$ which come before it, and not on the even-numbered terms. We shall therefore consider the two sequences

$$f_2, f_4, f_6, f_8, f_{10}, \ldots \quad \text{and} \quad f_1, f_3, f_5, f_7, f_9, \ldots$$

of even- and odd-numbered terms separately. The exercises which follow are designed to help you to find a formula for the nth term of the sequence $f_2, f_4, f_6, f_8, f_{10}, \ldots$ of even-numbered terms. [*Beware!* The nth term of the sequence $f_2, f_4, f_6, \ldots$ is f_{2n}, and not f_n.] I leave you to find a similar formula for the nth term of the odd-numbered sequence $f_1, f_3, f_5, \ldots$ (Exercise 57).

The nth term f_{2n} of the even-numbered sequence $f_2, f_4, f_6, \ldots$ satisfies the recurrence relation

$$f_{2n} = 2f_{2(n-1)} - f_{2(n-2)} + f_{2(n-4)}.$$

In other words, the nth, $(n-1)^{\text{th}}$, $(n-2)^{\text{th}}$, and $(n-4)^{\text{th}}$ terms of this sequence satisfy the recurrence relation

$$u_n = 2u_{n-1} - u_{n-2} + u_{n-4}. \tag{10.2}$$

There are in fact infinitely many different sequences all of which satisfy the recurrence relation (10.2). But each such sequence is completely determined once its first four terms u_1, u_2, u_3, u_4 have been specified.

We are looking for a formula which expresses the nth term f_{2n} of our sequence in terms of n alone. We have already seen how to get such a formula for the nth term F_n of the Fibonacci sequence by choosing a suitable combination of the two sequences of powers 1, a, a^2, a^3, ... and 1, b, b^2, b^3, ... which satisfy the recurrence relation (10.1). So if we want a formula for f_{2n} in terms of n alone it looks as though the first thing we should do is to find out which sequences of powers 1, x, x^2, x^3, x^4, ... satisfy the recurrence relation (10.2).

To do this properly we shall have to solve a quadratic equation whose roots are *complex* numbers. If you already know how to solve quadratic equations with complex roots and how to add and multiply complex numbers, then you should skip the next bit in italics and go straight on to Exercise 51.

[If we stick to ordinary real numbers, then we can certainly talk about 'the square root of 4' and 'the square root of 2'. But it makes no sense to talk about 'the square root of -3' or 'the square root of -1', because -3 and -1 are not the square of anything. So if we stick to ordinary real numbers we can solve equations like $x^2 = 4$ and $x^2 = 2$, but we have to say that equations like $x^2 = -3$ and $x^2 = -1$ 'have no solution'.

This turns out to be mathematically inconvenient. More than four hundred years ago mathematicians began to feel the urge to invent a new number $\sqrt{-1}$ whose square was equal to -1: $(\sqrt{-1})^2 = -1$. For a long time no-one was quite sure whether this new 'number' $\sqrt{-1}$ really deserved to be treated as a proper number, so it was called an 'imaginary' number.

This single new number $\sqrt{-1}$ whose square is equal to -1 then gives birth to lots of other new numbers such as

(i) $2 \times \sqrt{-1}$ whose square is $(2 \times \sqrt{-1})^2 = 2^2 \times (\sqrt{-1})^2 = 4 \times (-1) = -4$,

(ii) $\sqrt{3} \times \sqrt{-1}$ whose square is $(\sqrt{3} \times \sqrt{-1})^2 = (\sqrt{3})^2 \times (\sqrt{-1})^2 = 3 \times (-1) = -3$, and

(iii) $1 + \sqrt{-1}$ whose square can be easily calculated if one assumes that the familiar rule $(p + q)^2 = p^2 + 2pq + q^2$ works for these new

numbers:

$$(1+\sqrt{-1})^2 = 1^2 + 2 \times \sqrt{-1} + (\sqrt{-1})^2 = 1 + 2\sqrt{-1} + (-1) = 2 \times \sqrt{-1}.$$

These combinations of real and imaginary numbers are called 'complex' numbers.

One advantage of inventing complex numbers is that every quadratic equation $ax^2 + bx + c = 0$ can now be completely solved. For example, the equation $x^2 - x + 1 = 0$ has no ordinary real root. If we try to use the familiar formula for the roots of a quadratic equation, then we get

$$x = (1 + \sqrt{(1-4)})/2 = (1 + \sqrt{-3})/2$$

or

$$x = (1 - \sqrt{(1-4)})/2 = (1 - \sqrt{-3})/2.$$

If we stick to ordinary real numbers, then we have to reject these two numbers since they include the meaningless symbol $\sqrt{-3}$. But if we interpret $\sqrt{-3}$ as $\sqrt{3} \times \sqrt{-1}$, then we see that the familiar formula for the roots of a quadratic equation has in fact produced the two 'complex'.numbers $\frac{1}{2} \pm \frac{1}{2}\sqrt{-3}$. You should now check that each of these complex numbers is in fact a root of the quadratic equation $x^2 - x + 1 = 0$: for example,

$$(\tfrac{1}{2} + \tfrac{1}{2}\sqrt{-3})^2 = (\tfrac{1}{2})^2 + 2 \times \tfrac{1}{2} \times (\tfrac{1}{2}\sqrt{-3}) + (\tfrac{1}{2}\sqrt{-3})^2$$
$$= \tfrac{1}{4} + \tfrac{1}{2}\sqrt{-3} + (-\tfrac{3}{4}) = -\tfrac{1}{2} + \tfrac{1}{2}\sqrt{-3}$$
$$\therefore \quad (\tfrac{1}{2} + \tfrac{1}{2}\sqrt{-3})^2 - (\tfrac{1}{2} + \tfrac{1}{2}\sqrt{-3}) + 1 = (-\tfrac{1}{2} + \tfrac{1}{2}\sqrt{-3}) - (\tfrac{1}{2} + \tfrac{1}{2}\sqrt{-3}) + 1$$
$$= 0.$$

All I have tried to do here is to show how to do simple calculations with complex numbers. You will certainly want to know much more about them in due course. But if you are willing to postpone this more detailed study till later, then this short introduction to complex numbers should be enough to get you through the final stages of this investigation.]

Exercise 51 Suppose $1, x, x^2, x^3, x^4, \ldots$ is a sequence which satisfies the recurrence relation (10.2).
  (i) Show that $x^4 = 2x^3 - x^2 + 1$.
  (ii) Factorise $x^4 - 2x^3 + x^2 - 1$ as a product of two quadratic factors.
  (iii) Hence find the four possible values for the number x.

Exercise 52 Let $c = \tfrac{1}{2} + \tfrac{1}{2}\sqrt{-3}$ and $d = \tfrac{1}{2} - \tfrac{1}{2}\sqrt{-3}$.
  (i) Show that $c^2 = -d$, and that $cd = 1$.
  (ii) Show that $c^3 = d^3 = -1$.

(iii) Show that $1 + c - c^2 = 1 + d - d^2 = 2$.

(iv) Show that $c + c^2 = -(d + d^2)$.

The four possible values of x which you found in Exercise 51(iii) should have been your old friends $a = (1 + \sqrt{5})/2$, $b = (1 - \sqrt{5})/2$, and the two complex numbers c, d. Hence the four sequences

$$1, \ a, \ a^2, \ a^3, \ a^4, \ a^5, \ldots ,$$
$$1, \ b, \ b^2, \ b^3, \ b^4, \ b^5, \ldots ,$$
$$1, \ c, \ c^2, \ c^3, \ c^4, \ c^5, \ldots ,$$
$$1, \ d, \ d^2, \ d^3, \ d^4, \ d^5, \ldots ,$$

all satisfy the recurrence relation (10.2). If you multiply each term of the first sequence by some fixed number w, each term of the second sequence by some fixed number x, each term of the third sequence by some fixed number y, and each term of the fourth sequence by some fixed number z, you get four new sequences satisfying (10.2), namely

$$w, \ wa, \ wa^2, \ wa^3, \ wa^4, \ wa^5, \ldots ,$$
$$x, \ xb, \ xb^2, \ xb^3, \ xb^4, \ xb^5, \ldots ,$$
$$y, \ yc, \ yc^2, \ yc^3, \ yc^4, \ yc^5, \ldots ,$$
$$z, \ zd, \ zd^2, \ zd^3, \ zd^4, \ zd^5, \ldots .$$

If you now add together the corresponding terms of these four new sequences, you will get another sequence which also satisfies (10.2), namely

$$w + x + y + z, \ wa + xb + yc + zd, \ wa^2 + xb^2 + yc^2 + zd^2,$$
$$wa^3 + xb^3 + yc^3 + zd^3, \ldots .$$

By varying the values you choose for the fixed numbers w, x, y, and z you can produce infinitely many different sequences, each of which will satisfy the recurrence relation (10.2), and each of which is completely determined by its first four terms. So one will get in this way a sequence which is identical to our sequence

$$f_2, \ f_4, \ f_6, \ f_8, \ f_{10}, \ \ldots$$

precisely when w, x, y, z are chosen in such a way that the first four terms $w + x + y + z$, $wa + xb + yc + zd$, $wa^2 + xb^2 + yc^2 + zd^2$, $wa^3 + yc^3 + zd^3$ are equal to f_2, f_4, f_6, f_8 respectively. Hence w, x, y, and z

must be chosen to satisfy the equations

$$w \quad + x \quad + y \quad + z \quad = f_2 = 0,$$
$$wa + xb + yc + zd = f_4 = 1,$$
$$wa^2 + xb^2 + yc^2 + zd^2 = f_6 = 1,$$
$$wa^3 + xb^3 + yc^3 + zd^3 = f_8 = 2.$$

As soon as you manage to find numbers w, x, y, z satisfying these four equations you will have found the required formula expressing the number of 9-flips with $2n$ digits in terms of n along, namely

$$f_{2n} = wa^{n-1} + xb^{n-1} + yc^{n-1} + zd^{n-1}.$$

Exercise 53* (i) Use the known values of a, b, c, and d and the relations between c and d which you proved in Exercise 52, to find values of w, x, y, z which satisfy the four equations above. (ii) Hence write down a formula for f_{2n} in terms of n alone.

Hints to Exercises

47. (ii) (g) Show that $1 + a = a^2$. Then use this over and over again.
48. The third term x^2 must equal the sum $1 + x$ of the first two terms.
53. (i) Add the first two equations and subtract the third from the result. What do you know about $1 + a - a^2$ and $1 + b - b^2$? Look back at Exercise 52(iii). What does this now tell you about the value of $y + z$? Next use Exercise 52(ii) to simplify the fourth equation. Finally add the second and third equations together and subtract the simplified version of the fourth equation from the result. What do you know about $a + a^2 - a^3$ and $b + b^2 - b^3$? Look back at Exercise 52(iv). What does this tell you about the value of $y - z$?

Solutions

Exercises 46 233, 377, 610, 987, 1597.

Exercises 47 (i) (a) 0, 0, 0. (b) 8, 13, 21. (c) 5, 8, 13. (d) 29, 47, 76. (e) −13, −21, −34. (f) 16, 26, 42.

(ii) a, $b = (1 \pm \sqrt{5})/2$ happen to be the two solutions of $1 + x = x^2$, so $1 + a = a^2$ and $1 + b = b^2$. Multiplying these two equations by a^{n-1} and b^{n-1} respectively we see that $a^{n-1} + a^n = a^{n+1}$ and $b^{n-1} + b^n = b^{n+1}$ for every $n \geq 1$, so both sequences (g) and (h) satisfy (10.1).

Exercise 48 The third term must be the sum of the first two terms, so $1 + x = x^2$. Hence $x = (1 \pm \sqrt{5})/2$, so $x = a$ or $x = b$.

Exercise 49 (i) $u_n = u_{n-1} + u_{n-2}$, so $cu_n = cu_{n-1} + cu_{n-2}$.

(ii) $u_n = u_{n-1} + u_{n-2}$ and $v_n = v_{n-1} + v_{n-2}$, so $(u_n + v_n) = (u_{n-1} + v_{n-1}) + (u_{n-2} + v_{n-2})$.

Exercise 50 (i) $x + y = 0$, so $x = -y$. Substituting this in $xa + yb = 1$ we get $x(a - b) = 1$, so $x = 1/\sqrt{5}$, $y = -1/\sqrt{5}$.

(ii) $F_n = \dfrac{1}{\sqrt{5}} \left(\dfrac{1 + \sqrt{5}}{2} \right)^{n-1} - \dfrac{1}{\sqrt{5}} \left(\dfrac{1 - \sqrt{5}}{2} \right)^{n-1}$.

Exercise 51 (i) The fifth term must equal 'twice the fourth term minus the third term plus the first term', so $x^4 = 2x^3 - x^2 + 1$.

(ii) $x^4 - 2x^3 + x^2 - 1 = (x^2 - x - 1)(x^2 - x + 1)$.

(iii) $x^4 - 2x^3 + x^2 - 1 = 0$ by (i). But then either $x^2 - x - 1 = 0$ or $x^2 - x + 1 = 0$ by (ii). If $x^2 - x - 1 = 0$, then $x = a$ or $x = b$; if $x^2 - x + 1 = 0$, then $x = \frac{1}{2} + \frac{1}{2}\sqrt{-3}$ or $x = \frac{1}{2} - \frac{1}{2}\sqrt{-3}$.

Exercise 52 (i) $c^2 = (\frac{1}{2} + \frac{1}{2}\sqrt{-3})^2 = -\frac{1}{2} + \frac{1}{2}\sqrt{-3} = -d$. $cd = (\frac{1}{2} + \frac{1}{2}\sqrt{-3}) \times (\frac{1}{2} - \sqrt{-3}) = (\frac{1}{2})^2 - (\frac{1}{2}\sqrt{-3})^2 = 1$.

(ii) $c^2 = -d$, so $c^3 = -cd = -1$. Similarly $d^2 = (\frac{1}{2} - \frac{1}{2}\sqrt{-3})^2 = -\frac{1}{2} - \frac{1}{2}\sqrt{-3} = c$, so $d^3 = -cd = -1$.

(iii) $1 + c - c^2 = 1 + c + d = 2$. $1 + d - d^2 = 1 + d + c = 2$.

(iv) $c + c^2 = c - d = -(d - c) = -(d + d^2)$.

Exercise 53 (i) The hint shows that $y + z = 0$. Since $c^3 = d^3 = -1$, the fourth equation becomes $wa^3 + xb^3 = 2$. Adding the second and third equations and subtracting the simplified version of the fourth equation we get $(c + c^2)y + (d + d^2)z = 0$; so since $c + c^2 = -(d + d^2) \neq 0$, $y - z = 0$. Therefore $y = z = 0$ and the four original equations boil down to just two, namely $w + x = 0$ and $wa + xb = 1$. But these are precisely the equations we solved in Exercise 50(i), so $w = 1/\sqrt{5}$, $x = -1/\sqrt{5}$, $y = 0$, $z = 0$.

(ii) $f_{2n} = wa^{n-1} + xb^{n-1} + yc^{n-1} + zd^{n-1} = \dfrac{1}{\sqrt{5}} \left(\dfrac{1 + \sqrt{5}}{2} \right)^{n-1} - \dfrac{1}{\sqrt{5}} \left(\dfrac{1 - \sqrt{5}}{2} \right)^{n-1}$.

11 The end of the road

*When a person has discovered the truth about something
and has established it with great effort, then, on
viewing his discoveries more carefully, he often realises
that what he has taken such pains to find might have
been perceived with the greatest of ease.*
(Galileo, *On motion*)

When doing mathematics it is not at all unusual to discover, after
having done a considerable amount of work, some surprising fact
which would have made everything much, much simpler if only you
had noticed it earlier. But just in case your solution to Exercise 53 did
not quite reveal all, you had better work through the next exercise.

Exercise 54 (i) Use the recurrence relation $f_{2n} = 2f_{2(n-1)} - f_{2(n-2)} + f_{2(n-4)}$ to complete the table in Fig. 11.1. Where have you met this
sequence before?

n	2	4	6	8	10	12	14	16	18	20	22	24	26	28	30
f_{2n}	0	1	1	2											

Fig. 11.1

(ii) Now compare the formula for f_{2n} which you obtained in
Exercise 53(ii) with the formula for F_n which you obtained in Exercise
50(ii). What do you notice?

All this suggests rather strongly that the sequence $f_2, f_4, f_6, \ldots$
satisfies a much simpler, and much more familiar recurrence relation
than (10.2). In fact your solutions to Exercise 50(ii) and Exercise
53(ii) prove that $f_{2n} = F_n$.

So! Right at the very end of your search you have discovered and
proved that the sequence $f_2, f_4, f_6, \ldots$ is exactly the same as the
Fibonacci sequence $F_1, F_2, F_3, \ldots$. But now that you have discovered
this very simple fact in such a roundabout way, it is natural to look for
a simpler and much more direct explanation.

Exercise 55 Suppose you wanted to convince someone else that the number of 9-flips with $2n$ digits is exactly equal to the sum of the number of 9-flips with $2(n-1)$ digits and the number of 9-flips with $2(n-2)$ digits. Try to find a *simple and direct* explanation of this fact.

So after all your hard work you have finally discovered that Question *A* has a very simple answer with a very short proof. It would certainly have saved you a lot of effort if Exercise 55 had appeared way back at the beginning of Chapter 9. But you would have missed out on the whole experience of fumbling your way towards your own answer. *Mathematics is a human activity. It is a method of inquiry* (and an extraordinarily effective one too). So no matter how important it may be for you to study the mathematics which other people have discovered, **if you want to understand what mathematics is really about, you have to join in the time-consuming search for your own results. There are no short cuts.**
The formula

$$f_{2n} = \frac{1}{\sqrt{5}} (a^{n-1} - b^{n-1})$$

which you found in Exercise 53(ii) can in fact be simplified a bit more.

Exercise 56 (i) Use a calculator to fill in the table in Fig. 11.2.

n	1	2	3	4	5	6	7	8	9	10	11	12	13	14	15
f_{2n}															
$a^{n-1}/\sqrt{5}$															
$b^{n-1}/\sqrt{5}$															

Fig. 11.2

(ii) Show that f_{2n} is the nearest whole number to $a^{n-1}/\sqrt{5}$, for every $n \geq 1$.

(iii) How close are the two numbers $a^{n-1}/\sqrt{5}$ and f_{2n} in practice? Use your calculator to work out the difference

$$b^{n-1}/\sqrt{5} = (a^{n-1}/\sqrt{5}) - f_{2n}$$

when $n = 37$.

It is now your turn to complete this investigation by finding a formula which expresses the $(n + 1)th$ *term* f_{2n+1} of the odd-numbered sequence $f_1, f_3, f_5, \ldots$ in terms of n alone.

Exercise 57   Find a formula for f_{2n+1} in terms of n alone.

I shall stop at this point. But there are still some unanswered questions you might think about. For example,

Question *E*   *Is there some connection between Exercise 1 and Exercise 2? Why does the same number crop up in both places?*

Hints to Exercises

55. 9-flips with $2n$ digits are of two types. The first type consists of '*straddlers*', which have a central block straddling the midpoint (between the nth and $(n + 1)^{th}$ digits). The second type consists of '*non-straddlers*', which split into two halves. Consider these two types separately.

Solutions

Exercise 54   (i) The sequence $f_2, f_4, f_6, \ldots$ is the same as the Fibonacci sequence $F_1, F_2, F_3, \ldots$.
(ii) The two formulae are identical.

Exercise 55   Use the hint. Let S_{2n} = the number of *straddlers* with $2n$ digits, and N_{2n} = the number of *non-straddlers* with $2n$ digits. Then $f_{2n} = S_{2n} + N_{2n}$. Now S_{2n} = (number of straddlers whose central block has length $=4$) + (number of straddlers whose central block has length >4). If the central block has length $=4$, then the central block is 1089, and when we remove this block we get a non-straddler with $2n - 4$ digits: so the first bracket is equal to N_{2n-4}. If the central block has length >4, then it must have an even number of 9s in the middle, and when we remove the two central 9s we get a straddler with $2n - 2$ digits: so the second bracket is equal to S_{2n-2}. Therefore $S_{2n} = N_{2n-4} + S_{2n-2}$. In the same spirit we observe that N_{2n} = (number of non-straddlers whose middle two digits are 00) + (number of non-straddlers whose middle four digits are 8910). If the middle two digits are 00, then we may remove these two zeros to get a non-straddler with $2n - 2$ digits: so the first bracket is equal to N_{2n-2}. If the middle four digits are 8910, then we may remove these four digits to get a straddler with $2n - 4$ digits: so the second bracket is equal to S_{2n-4}. Therefore $N_{2n} = N_{2n-2} + S_{2n-4}$. Hence

$$f_{2n} = S_{2n} + N_{2n} = (N_{2n-4} + S_{2n-2}) + (N_{2n-2} + S_{2n-4})$$
$$= (N_{2n-4} + S_{2n-4}) + (N_{2n-2} + S_{2n-2}) = f_{2n-4} + f_{2n-2}.$$

Exercise 56 (i)

n	1	2	3	4	5	6	7	8
f_{2n}	0	1	1	2	3	5	8	13
$a^{n-1}/\sqrt{5}$	0.4472	0.7236	1.1708	1.8944	3.0652	4.9597	8.0249	12.9846
$b^{n-1}/\sqrt{5}$	0.4472	−0.2764	0.1708	−0.1056	0.0652	−0.0403	0.0249	−0.0154

n	9	10	11	12	13	14	15
f_{2n}	21	34	55	89	144	233	377
$a^{n-1}/\sqrt{5}$	21.0095	33.9941	55.0036	88.9978	144.0014	232.9991	377.0005
$b^{n-1}/\sqrt{5}$	0.0095	−0.0059	0.0036	−0.0022	0.0014	−0.0009	0.0005

(ii) $f_{2n} = a^{n-1}/\sqrt{5} - b^{n-1}/\sqrt{5}$ is always a whole number. $|b^{n-1}/\sqrt{5}|$ is the distance between $a^{n-1}/\sqrt{5}$ and this whole number; and this distance is always less than $\frac{1}{2}$, since $\frac{1}{2} > |b^0/\sqrt{5}| > |b/\sqrt{5}| > |b^2/\sqrt{5}| > \ldots$.

(iii) As far as your calculator is concerned this number is probably indistinguishable from zero.

Exercise 57 The sequence $f_3, f_5, f_7, \ldots$ whose n^{th} term is f_{2n+1} satisfies exactly the same recurrence relation (10.2) as the sequence $f_2, f_4, f_6, \ldots$ whose n^{th} term is f_{2n}. Each sequence which satisfies (10.2) is completely determined as soon as we know its first four terms. But the first four terms of the odd-numbered sequence ($f_3 = 0, f_5 = 1, f_7 = 1, f_9 = 2$) are exactly the same as the first four terms of the even-numbered sequence ($f_2 = 0, f_4 = 1, f_6 = 1, f_8 = 2$), so the two sequences are identical: that is, $f_{2n+1} = f_{2n}$ for every $n \geq 1$. (*Alternatively*, a 9-flip with $2n + 1$ digits has either a 9 or a 0 in its central position. Removing this central digit produces a 9-flip with $2n$ digits. Therefore $f_{2n+1} = f_{2n}$ for every $n \geq 1$.)

Postscript

Even when you have found what looks like a satisfactory solution to a mathematical problem, you can be sure that there are plenty of other ways of thinking about the same question. And some of these other approaches will probably be better than the approach you first chose. After I had completed the first draft of this extended investigation I thought I understood the problem fairly well. But I was wrong. Some weeks later I received the following letter from Graham McCauley, and I am grateful for his permission to reproduce it here.

Tony,

I've just spent an enjoyable evening reading (and working through) Flips. You may be interested to know that you quite successfully led me to a slightly different solution from the one you produced. Here is what happened.

Exercises 1 and 2 in Chapter 2 got me thinking very much in terms of complements to cope with different bases. Using a 'bar' notation reminiscent of old fashioned logarithms I emerged from Exercise 5 in Chapter 2 ready to write 1089 as $11\bar{1}\bar{1}$ (which is a 11-flip in any base $b \geq 2$).

Unfortunately I abandoned this notation for the base 10 work [in Chapter 5, Chapter 6 and the first half of Chapter 7] till I reached Exercise 21. It took me about five minutes to get hold of 2178, but once I spotted that it was $22\bar{2}\bar{2}$ I was on to a winning streak up to Exercise 33.

I followed you diligently up to Exercise 45 (well, fairly diligently, as I didn't need to check the shortcomings of programmable calculators!). I found Exercise 45 quite hard and so became restive for the question that finally showed up as Exercise 54. In fact the stuff in between is quite standard so I settled down and eventually came up with the following.

Flips are 'antisymmetric' strings of double ones, zeros, and double one-bars, with the pairs of ones and one-bars occurring alternately:

e.g. $11\bar{1}\bar{1}$, $110\bar{1}\bar{1}011\bar{1}\bar{1}$, $1100\bar{1}\bar{1}\bar{1}\bar{1}0\bar{1}\bar{1}011\bar{1}\bar{1}011011\bar{1}\bar{1}00\bar{1}\bar{1}$.

A $(2n + 1)$-digit flip has a central zero, which can be removed to give a $2n$-digit flip, so $f_{2n+1} = f_{2n}$ as we have seen. A $2n$-digit flip with two (or more) central zeros can have one pair of zeros removed to give a

$(2n - 2)$-*digit flip*; *otherwise the central sequence is either* $11\overline{1}\overline{1}$ *or* $\overline{1}111$
and can be removed to leave a $(2n - 4)$-*digit flip. Hence* $f_{2n} = f_{2n-2} +$
f_{2n-4}, *without the need to separate straddlers and non-straddlers.*
Viewed from the other end the enumeration is like the heifer
population: $11\overline{1}\overline{1}$ *ages to* $1100\overline{1}\overline{1}$, *and then to* $110000\overline{1}\overline{1}$, *at which stage*
she is fertile and produces one offspring, namely $11\overline{1}\overline{1}11\overline{1}\overline{1}$. *The great*
progenitor produces one new offspring each year, and each offspring
becomes fertile in its second year. I guess f_{2n+1} *is the winter population*
in this model!

Graham.

Investigation II

The postage stamp problem

12 Getting stuck in

We start with an old puzzle.

Problem   You have an 8 pint jug, a 5 pint jug, and a tap. Show how to obtain exactly 1 pint.

The challenge is to obtain a specified quantity (1 pint) in terms of other quantities (5 pints and 8 pints). One solution can be expressed by the equation $1 = (8 + 8) - (5 + 5 + 5)$. In other words, fill the 8 pint jug twice over $(8 + 8)$, transferring its contents each time to fill the 5 pint jug, whose contents we then *throw away* $(-(5 + 5 + 5))$. This uses both addition and *subtraction*. The starting point for the present investigation is slightly different. Once again you have to combine whole numbers to obtain other whole numbers. But this time *you are not allowed to subtract*.

Exercise 1*   You have an inexhaustible supply of 5 cent and 8 cent stamps.
   (i) Which of these amounts can be made using 5c and 8c stamps only?

   23c, 42c, 76c, 951c, 13632c, 27c, 17c, 11c, 14c, 26c, 28c.

   (ii) Make a complete list of the amounts between 1c and 99c which cannot be made.

Exercise 2*   You have an inexhaustible supply of 5c and 13c stamps.
   (i) Which of these amounts can be made using 5c and 13c stamps only?

   23c, 24c, 25c, 26c, 73c, 872c, 351642c,

   17c, 84c, 36c, 37c, 47c, 48c, 49c.

   (ii) Make a complete list of the amounts between 1c and 99c which cannot be made.

Exercise 1 and Exercise 2 are particular examples of a single mathematical problem.

Problem   You have an inexhaustible supply of two kinds of stamps—say a cent and b cent. Which amounts can be made using just these stamps? Which amounts cannot be made?

This is the problem which you are invited to explore in the rest of this investigation.

You would obviously like to know how to predict which amounts can be made, and which cannot be made, *as soon as you are told the values (a and b) of the two kinds of stamps which are being used.* There must surely be some connection between the two numbers a and b and the amounts which can be obtained by combining a and b. But at this stage it is not at all clear what sort of connection to look for! One way of getting your teeth into a problem like this is

to experiment,

and

to record your results in a systematic way,

in the hope that you might begin to see some kind of pattern. So perhaps you should go right ahead and choose some values for a and b and just see what happens! But once you decide to experiment like this, you should always try to choose the values of a and b *in a systematic way*—otherwise you will find it exceedingly difficult to spot any pattern at all. So to start with **we shall keep $a = 5$ fixed, and try first $b = 2$, then $b = 3$, then $b = 4$, and so on.**

Exercise 3*   Complete and extend the table in Fig. 12.1 until you think you can say exactly which amounts *can* be made using 5c and 2c stamps only, and which amounts *cannot* be made.

Total amount	0	1	2	3	4	5	6	7	8	9	
Can be made	√		√								
Cannot be made		×									

Fig. 12.1

Exercise 4*   Complete and extend the table in Fig. 12.2 until you think you can say exactly which amounts *can* be made using 5c and 3c stamps only, and which amounts *cannot* be made.

Total amount	0	1	2	3	4	5	6	7	8	9	10	
Can be made	√			√								
Cannot be made		×	×									

Fig. 12.2

It is a bit awkward having to refer over and over again to 'amounts which can be made' and 'amounts which cannot be made'. If *a* cent and *b* cent are the two values of stamps being used, then we shall call a number
  good if it can be obtained by combining the two numbers *a* and *b,* and
  bad if it cannot be obtained by combining the two numbers *a* and *b.*
So in Exercise 3 the numbers 4 and 6 are good. But in Exercise 4 the number 4 is bad and the number 6 is good.

Exercise 5* Suppose you have only 5c and 4c stamps. Draw up a table like the one in Fig. 12.3, and extend it until you think you can say exactly which numbers are *good* and which are *bad.*

Number	0	1	2	3	4	5	6	7	8	9	10	11	12
Good?													
Bad?													

Fig. 12.3

Exercise 6* Suppose you have only 5c and 6c stamps. Draw up a table like the one in Exercise 5 and extend it until you think you can say exactly which numbers are good and which are bad.

Exercise 7* Suppose you have only 5c and 7c stamps. Draw up a table like the one in Exercise 5 and extend it until you think you can say exactly which numbers are good and which are bad.

You should by now have a fairly good idea of what is likely to happen when you have 5c and 8c stamps only, or 5c and 9c stamps only. *For in every case so far there appears to be a **cut-off point** where the bad numbers seem to come to an end.* At this stage the idea that there might *always* be such a cut-off point is no more than a hopeful guess. But it is an interesting guess, and is certainly worth following up.

The first thing to do with an interesting guess like this is to *test it.* And the most simple-minded way of testing this particular guess is to see whether it works for values of *a* and *b* which you haven't tried before. If your guess survives this initial test, you should begin looking for some convincing mathematical *explanation,* or *proof,* which would show that your guess is in fact correct.

So let's start by testing this guess that there will always be a cut-off point where the bad numbers seem to come to an end.

Exercise 8* (i) Use your solutions to Exercises 3–7 to complete the table in Fig. 12.4

a	5	5	5	·	5	5	
b	2	3	4	·	6	7	
Apparent cut-off point (if there is one)	3/4	7/8					

Fig. 12.4

(ii) For each of the pairs *a*, *b* below draw up a table like the one in Exercise 5.

(a) *a* = 5, *b* = 8; (b) *a* = 5, *b* = 9; (c) *a* = 5, *b* = 10;
(d) *a* = 5, *b* = 11; (e) *a* = 5, *b* = 12; (f) *a* = 5, *b* = 13.

(iii) Use your solution to Part (ii) to extend the table in Fig. 12.4 as far as *b* = 13.

Before jumping to too many conclusions you should obviously test this cut-off point idea for values of *a* other than *a* = 5. But it is beginning to look as though the idea is more or less correct, at least when *a* = 5. Among all the pairs *a*, *b* you have looked at so far, *the only pair for which there was no cut-off point at all was the pair a = 5, b = 10.* This apparent exception should not really surprise you.

Because if *a* and *b* are both multiples of 5, then any number which we get by combining *a* and *b* is bound to be a multiple of 5. But then numbers which are not multiples of 5 will always be bad, so the bad numbers will never come to an end.

You may have already tried to spot some kind of pattern in the list of apparent cut-off points which you compiled in Exercise 8. If you did, then you may have noticed that your list suggests an improved version of the original cut-off point idea. For it really looks as though you can actually guess *where the cut-off point comes,* at least when *a* = 5.

Exercise 9* (i) Look back at the list of apparent cut-off points which you compiled in Exercise 8. Try to find a pattern which would help you *guess* the likely cut-off point when *a* = 5, *b* = 14.

(ii) Check your guess by drawing up a table like the one in Exercise 5 for *a* = 5, *b* = 14.

(iii) Guess the likely cut-off point when *a* = 5, *b* = 15. Then check whether your guess was correct.

(iv) Guess the likely cut-off point when *a* = 5, *b* = 16. Then check whether your guess was correct.

Exercise 10* (i) Suppose *a* = 5 and that *b* is some number which is not a multiple of 5. Guess a *formula* which gives the cut-off point in terms of *b*. Then check whether your formula really works.

(ii) Suppose *a* is any old number (not necessarily *a* = 5). Can you guess a formula which gives the cut-off point in terms of *a* and *b*? If you can guess such a formula, check whether it works for these pairs.

(*a*) *a* = 6, *b* = 7; (*b*) *a* = 6, *b* = 8; (*c*) *a* = 6, *b* = 9.

Hints to Exercises

9. (i) 3/4, 7/8, 11/12, - - -, 19/20, 23/24, 27/28, 31/32, - - -, 39/40, 43/44, ...

Solutions

Exercise 1 (i) 23, 42, 76, 951, 13632, 26, 28. (27c, 17c, 11c, 14c cannot be made.)

(ii) 1, 2, 3, 4, 6, 7, 9, 11, 12, 14, 17, 19, 22, 27.

Exercise 2 (i) 23, 25, 26, 73, 872, 351642, 84, 36, 48, 49. (24c, 17c, 37c, 47c cannot be made.)

(ii) 1, 2, 3, 4, 6, 7, 8, 9, 11, 12, 14, 16, 17, 19, 21, 22, 24, 27, 29, 32, 34, 37, 42, 47.

Exercise 3 It looks as though 1c and 3c are the only amounts which cannot be made.

Exercise 4 It looks as though 1c, 2c, 4c, and 7c are the only amounts which cannot be made.

Exercise 5 It looks as though 1, 2, 3, 6, 7, and 11 are the only bad numbers.

Exercise 6 It looks as though 1, 2, 3, 4, 7, 8, 9, 13, 14, and 19 are the only bad numbers.

Exercise 7 It looks as though 1, 2, 3, 4, 6, 8, 9, 11, 13, 16, 18, and 23 are the only bad numbers.

Exercise 8

a	5	5	5	–	5	5	5	5	–	5	5	5
b	2	3	4	–	6	7	8	9	–	11	12	13
Apparent cut-off point (if there is one)	3/4	7/8	11/12	–	19/20	23/24	27/28	31/32	–	39/40	43/44	47/48

Exercise 10 (i) *Guess*: Cut-off point between $4b - 5/4(b - 1)$.

(ii) *Guess*: Cut-off point between $ab - (a + b)/(a - 1)(b - 1)$. Testing seems to confirm this guess: for example, when $a = 4$ and $b = 7$ the apparent cut-off point occurs between $17/18 = 4 \times 7 - (4 + 7)/(4 - 1)(7 - 1)$.

13 Coming unstuck: the search for proof

But we are running ahead of ourselves! When $a = 5$ you now *believe* that you can predict exactly where the cut-off point comes for any value of b. You may even have a formula which you *believe* predicts where the cut-off point comes for any pair of numbers a, b. But how can you be sure that your formula always works? Up to now you haven't even explained **why there will always be a cut-off point**. And even if you managed to convince yourself that there will always be a cut-off point, you would still have to explain **why the cut-off point always occurs exactly where you think it does**. However now that you have something worth explaining, you can go back to the beginning in search of some kind of *mathematical explanation, or proof.*

Exercise 11* In Exercise 3 you had to decide exactly which numbers were good and which were bad when $a = 5$ and $b = 2$ (Fig. 13.1).

Number	0	1	2	3	4	5
Good?	√		√		√	√
Bad?		×		×		

Fig. 13.1

You could in fact have stopped as soon as you came across *two good numbers in a row,* namely

$$4 = 0 \times 5 + 2 \times 2 \quad \text{and} \quad 5 = 1 \times 5 + 0 \times 2.$$

Because as soon as you know that 4 and 5 are both good, it is easy to show that *each number* $\geqslant 4$ *is good* (so 3 must be the last bad number). Can you explain why?

Exercise 12* In Exercise 4 you had to decide exactly which numbers were good and which were bad when $a = 5$ and $b = 3$ (Fig. 13.2).
  (i) Write each of the numbers 8, 9, and 10 in the form $x \times 5 + y \times 3$.

Number	0	1	2	3	4	5	6	7	8	9	10
Good?	√			√		√	√		√	√	√
Bad?		×	×		×			×			

Fig. 13.2

(ii) In Exercise 4 you could in fact have stopped as soon as you came across *three good numbers in a row*, namely 8, 9, and 10. Because it is then easy to show that *each number* ≥ 8 *is good* (so 7 must be the last bad number). Can you explain how?

Exercise 13* (i) Suppose you had to decide exactly which numbers are good and which numbers are bad when $a = 5$ and $b = 4$. How many good numbers in a row would you have to find before you could stop? How soon could you have stopped in Exercise 5?

(ii) Suppose you had to decide exactly which numbers are good and which numbers are bad when $a = 5$ and $b = 6$. How many good numbers in a row would you have to find before you could stop? How soon could you have stopped in Exercise 6?

(iii) How soon could you have stopped in Exercise 7?

So if there really is a last bad number, you now have a way of finding it as soon as you know the values of a and b. For example, if $a = 5$ and b is any number which is not a multiple of 5 (say $b = 23$, or $b = 2233$, or ...), then you only have to check each of the numbers 0, 1, 2, 3, ... until you find *five consecutive good numbers*.

Exercise 14 (i) Show that when $a = 5$ and $b = 23$, the five consecutive numbers 115, 116, 117, 118, 119 are all good.

(ii) Part (i) shows that when $a = 5$ and $b = 23$, each number ≥ 115 is good. So there must be a *last bad number*, and this last bad number is ≤ 114. Find the last bad number.

This method is a considerable improvement on plain guesswork. But it is still a bit unsatisfactory. Even if a is kept fixed (say at $a = 5$), it looks as though the numbers you have to test are going to get uncomfortably large as the value of b gets bigger. Worse still, *you are not yet sure that there will always be some cut off point beyond which all numbers are good.* So for all you know at present you may occasionally land up testing larger, and larger numbers without ever finding five good numbers in a row. We shall therefore fix $a = 5$ for the moment and try to show that *one can always find five consecutive good numbers, no matter what value is chosen for b.*

Suppose we write the unspecified value of 'b' simply as b. It might then be possible somehow to use elementary algebra to write down five consecutive good numbers. In Exercise 10(i) you probably decided that, when $a = 5$, the cut-off point seems to occur between $4b - 5$ and $4b - 4$. So you would like to show that

$$4b - 5 \text{ is always bad, and}$$
$$4b - 4, \ 4b - 3, \ 4b - 2, \ 4b - 1, \ 4b \text{ are always good.}$$

Exercise 15 (i) Suppose $a = 5$ and $b = 21$. Show that $4b - 5$ is bad.
(ii) Suppose $a = 5$ and $b = 201$. Show that $4b - 5$ is bad.
(iii) Suppose $a = 5$ and $b = 2001$. Show that $4b - 5$ is bad.

Exercise 16* Suppose $a = 5$ and b is not a multiple of 5. Show that $4b - 5$ is always bad.

So all you have to do now is to show that, when $a = 5$ and b is not a multiple of 5, the five consecutive numbers $4b - 4$, $4b - 3$, $4b - 2$, $4b - 1$, $4b$ are all good. You could then be absolutely certain that all numbers $\geq 4b - 4$ are good. Exercise 16 would then guarantee that $4b - 5$ is always the last bad number.

Now the number $4b (= 0 \times 5 + 4 \times b)$ is obviously good, so you only have to worry about the four numbers $4b - 4$, $4b - 3$, $4b - 2$, $4b - 1$. To get started you will probably need the following very simple hint.

Hint: *Write b as a multiple of 5 plus a remainder $r \leq 4$: $b = 5q + r$.*
Since b is not a multiple of 5 you will never get $r = 0$. So there are just four possible values of r: namely $r = 1$, $r = 2$, $r = 3$, or $r = 4$. The next exercise considers each of these possible values of r in turn. You will have to use some elementary algebra to solve it, but if you get stuck don't be afraid to use the hint at the end of the chapter.

Exercise 17 (i) Suppose that $r = 1$, so that $b = 5q + 1$.
(*a*) Show that $4b - 4$ is good. (*b*) Show that $4b - 3$ is good.
(*c*) Show that $4b - 2$ is good. (*d*) Show that $4b - 1$ is good.
(ii) Suppose that $r = 2$, so that $b = 5q + 2$.
(*a*) Show that $4b - 3$ is good. (*b*) Show that $4b - 1$ is good.
(*c*) Show that $4b - 4$ is good. (*d*) Show that $4b - 2$ is good.

(iii) Suppose that $r = 3$, so that $b = 5q + 3$.
(a) Show that $4b - 2$ is good. (b) Show that $4b - 4$ is good.
(c) Show that $4b - 1$ is good. (d) Show that $4b - 3$ is good.
(iv) Suppose that $r = 4$, so that $b = 5q + 4$.
(a) Show that $4b - 1$ is good. (b) Show that $4b - 2$ is good.
(c) Show that $4b - 3$ is good. (d) Show that $4b - 4$ is good.

Your solutions to Exercises 16 and 17 now show that the guess you made way back in Exercise 10(i) is in fact true: *when a = 5 and b is any number which is not a multiple of 5, then 4b − 5 is the last bad number.*

The original decision to fix the value of $a = 5$ was part of our attempt to investigate the main problem in a systematic way. You carried out this systematic investigation of the case $a = 5$ in Exercises 1–8. And the fact that $a = 5$ was kept fixed made it much easier to spot the pattern in Exercise 10(i). It also meant that the algebra you needed in Exercises 16 and 17 was not too difficult. *But what if a is not equal to 5?* What if $a = 6$, or $a = 106$, or ... ? Will the method you used in Exercises 16 and 17 still work?

Exercise 18 Suppose $a = 6$.
(i) If b is a multiple of 2, show that there is no last bad number, and so no cut-off point at all.
(ii) If b is a multiple of 3, show that there is no last bad number, and so no cut-off point at all.

Exercise 19 Suppose $a = 6$ and b is an unspecified number which is neither a multiple of 2 nor a multiple of 3.
(i) Show that $5b - 6$ is always bad.
(ii) Write b as a multiple of 6 plus a remainder $r \leqslant 5$: $b = 6q + r$.
(a) Show that either $r = 1$ or $r = 5$.
(b) Show that when $r = 1$, the numbers $5b - 5$, $5b - 4$, $5b - 3$, $5b - 2$, $5b - 1$ are all good.
(c) Show that when $r = 5$, the numbers $5b - 5$, $5b - 4$, $5b - 3$, $5b - 2$, $5b - 1$ are all good.
(iii) Show that $5b - 6$ is always the last bad number.

So long as a is fixed it is beginning to look as if you might be able to use elementary algebra to prove that your guess is correct. But wait a minute! When $a = 5$ you wrote b as a multiple of 5 plus a remainder $r \leqslant 4$. And you then had to look at *each of the four possible remainders $r = 1$, $r = 2$, $r = 3$, $r = 4$ one at a time.* And for each of these four possible values of r you had to show that *each of the four numbers $4b - 1$, $4b - 2$, $4b - 3$, $4b - 4$ is good*. If you try to use exactly the same method when $a = 101$, for example, then you are going to have to write b as a multiple of 101 plus a remainder $r \leqslant 100$. You will then have to look at *each of the one hundred possible remainders $r = 1$, $r = 2$, . . . , $r = 100$ one at a time*. And for each of these one hundred possible values of r you will have to show that *each of the one hundred numbers $100b - 1$, $100b - 2$, $100b - 3$, . . . , $100b - 100$ is good*. It looks as if this is going to involve a ridiculous amount of work just to prove that your guess is true when $a = 101$. But perhaps the method is not quite as useless as the example $a = 101$ makes it look.

Suppose a and b are both unspecified numbers. If a and b are both even (that is, if a and b are both *multiples of 2*), then every good number $xa + yb$ will also be a *multiple of 2*. But then every odd number will be bad, so there can be no last bad number. Similarly if a and b are both *multiples of some number c*, then every good number $xa + yb$ will also be a *multiple of c*. But then every number which is not a multiple of c will be bad, so there can be no last bad number. So the guess you made in Exercise 10(iii) that '*every number $\geqslant (a - 1)(b - 1)$ is good and $ab - (a + b)$ is the last bad number*' cannot possibly be true unless a *and b have no common factors*. Thus if you want to prove that your guess is correct, then the proof must depend in some way on the fact that a and b have no common factors.

So here's you chance to generalize the method which worked for $a = 5$ (Exercises 16 and 17) and for $a = 6$ (Exercise 19) to prove that '$ab - (a + b)$ *is always the last bad number* (provided that a and b have no common factors)'. The algebra is more complicated than before, but see how far you can get. If you get stuck, there is a hint to help you. And if you get really stuck, go on to the next chapter: we shall come back to look at the same problem in a different way later on.

Project Suppose that a and b are positive whole numbers with no common factors, and that $a < b$.

(i) Show that $ab - (a + b)$ is always bad.

(ii) Write b as a multiple of a plus a remainder $r \leqslant a - 1$: $b = qa + r$.

    (*a*) Show that a and r have no common factors.

    (*b*) Show that, no matter what the value of r may be, the numbers

$$(a - 1)(b - 1), \ (a - 1)(b - 1) + 1, \ (a - 1)(b - 1) + 2, \ . . . , \ (a - 1)b$$

are all good.

(iii) Show that $ab - (a + b)$ is always the last bad number.

Hints to exercises

11. $4 = 0 \times 5 + 2 \times 2$ is good, and $5 = 1 \times 5 + 0 \times 2$ is good;
$6 = 4 + 2 = 0 \times 5 + 3 \times 2$ is good, and $7 = 5 + 2 = 1 \times 5 + 1 \times 2$ is good;
$8 = 6 + 2 = 0 \times 5 + 4 \times 2$ is good, and $9 = 7 + 2 = 1 \times 5 + 2 \times 2$ is good;
$10 = 8 + 2 = \ldots$ is good, and $11 = 9 + 2 = \ldots$ is good.

12. (ii) $8 = 1 \times 5 + 1 \times 3$ is good; $9 = 0 \times 5 + 3 \times 3$ is good; $10 = 2 \times 5 + 0 \times 3$ is good; $11 = 8 + 3 = 1 \times 5 + 2 \times 3$ is good; $12 = 9 + 3 = 0 \times 5 + 4 \times 3$ is good; $13 = 10 + 3 = 2 \times 5 + 1 \times 3$ is good; $14 = 11 + 3 = 1 \times 5 + 3 \times 3$ is good; $15 = 12 + 3 = 0 \times 5 + 5 \times 3$ is good; $16 = 13 + 3 = 2 \times 5 + 2 \times 3$ is good; and so on.

14. (ii) In Exercise 10(i) you guessed that the likely cut-off point occurs between $4(b - 1) - 1$ and $4(b - 1)$. You must now check, when $b = 23$, that
(a) $4(b - 1) - 1$ really is bad, and
(b) all numbers $\geqslant 4(b - 1) = 88$ are good.

15. (i) Suppose $4b - 5 = 79$ could be obtained by combining $a = 5$ and $b = 21$: say $79 = 5x + 21y$. How many 21s could you use? Try each of the four possible values of y in turn.

(ii) Suppose $4b - 5 = 799$ could be obtained by combining $a = 5$ and $b = 201$: say $799 = 5x + 201y$. How many 201s could you use? Try each of the four possible values of y in turn.

16. Suppose $4b - 5 = 5x + by$. What are the four possible values for y? Try each one in turn.

17. (i) (a) $4b - 4 = 4(5q + 1) - 4 = \ldots$ (b) $4b - 3 = 4(5q + 1) - 3 = ? \times (5q + 1) + ? \times 5$.

(ii) (a) $4b - 3 = 4(5q + 2) - 3 = \ldots$ (b) $4b - 1 = 4(5q + 2) - 1 = \ldots$

19. (i) Suppose $5b - 6 = 6x + by$. What are the five possible values for y? Try each one in turn.

(ii) (a) Show that if $r = 0$, 2, 3, or 4, then b is either a multiple of 2 or a multiple of 3.

(b), (c) Try to imitate your solution to Exercise 17, but this time with $a = 6$ instead of $a = 5$, and with just two possible values of r ($r = 1$ and $r = 5$) instead of four possible values of r.

Project (i) Suppose $ab - (a + b) = ax + by$ ($x, y \geqslant 0$). Show that $y \leqslant a - 2$. Why does this contradict $b(a - 1 - y) = a(x - 1)$?

(ii) (b) the number $(a - 1)b$ is obviously good. Your task is therefore to show that when k is any one of the numbers $a - 1$, $a - 2$, $a - 3$, $\ldots$, 1, the number $(a - 1)b - k = (a - 1)(qa + r) - k = (a - 1)qa + [(a - 1)r - k]$ is always good. One way of doing this is to show that the last bracket $[(a - 1)r - k]$ can always be written as a multiple of a plus a multiple of b.

Solutions

Exercise 11 $a = 5$ and $b = 2$. If n is good, then $n = 5x + 2y$ for some $x, y \geqslant 0$. But then $n + 2 = 5x + 2(y + 1)$, $n + 4 = 5x + 2(y + 2)$, $n + 6 = 5x + 2(y + 3)$, $\ldots$ are all good. If $n + 1$ is also good, then $n + 1 = 5u + 2v$ for some $u, v \geqslant 0$, so $n + 3 = 5u + 2(v + 1)$, $n + 5 = 5u + 2(v + 2)$, $n + 7 = 5u + 2(v +$

3), ... are all good. So if n and $n + 1$ are both good, then all numbers $\geqslant n$ are good.

Exercise 12   (i) $8 = 1 \times 5 + 1 \times 3$, $9 = 0 \times 5 + 3 \times 3$, $10 = 2 \times 5 + 0 \times 3$.

(ii) $a = 5$ and $b = 3$. If n is good, then $n = 5x + 3y$ for some $x, y \geqslant 0$. But then $n + 3 = 5x + 3(y + 1)$, $n + 6 = 5x + 3(y + 2)$, $n + 9 = \ldots$ are all good. If $n + 1$ is also good, then $n + 1 = 5u + 3v$ for some $u, v \geqslant 0$, so $n + 4 = 5u + 3(v + 1)$, $n + 7 = 5u + 3(v + 2)$, $n + 10 = \ldots$ are all good. If $n + 2$ is good as well, then $n + 2 = 5s + 3t$ for some $s, t \geqslant 0$, so $n + 5 = 5s + 3(t + 1)$, $n + 8 = 5s + 3(t + 2)$, $n + 11 = \ldots$ are all good. So if n, $n + 1$, and $n + 2$ are all good, then all numbers $\geqslant n$ are good.

Exercise 13   (i) $b = 4$ is the smaller of the two numbers a, b, so you only need to find *four* good numbers in a row. In Exercise 5 you could have stopped at 15 (because 12, 13, 14, and 15 are all good, so every number $\geqslant 12$ must be good).

(ii) $a = 5$ is the smaller of the two numbers a, b, so you only need to find *five* good numbers in a row. In Exercise 6 you could have stopped at 24 (because 20, 21, 22, 23, and 24 are all good, so every number $\geqslant 20$ must be good).

(iii) In Exercise 7 $a = 5$ is the smaller of the two numbers a, b, so you only needed to find *five* good numbers in a row. You could therefore have stopped at 28 (because 24, 25, 26, 27, and 28 are all good, so every number $\geqslant 24$ must be good).

Exercise 14   (i) $115 = 0 \times 5 + 5 \times 23$, $116 = 14 \times 5 + 2 \times 23$, $117 = 5 \times 5 + 4 \times 23$, $118 = 19 \times 5 + 1 \times 23$, $119 = 10 \times 5 + 3 \times 23$.

(ii) In Exercise 10(i) you guessed that the cut-off point would occur between $4b - 5 = 87$ and $4(b - 1) = 88$. To check that this guess is correct you have to do two things: (*a*) show that 87 is bad; (*b*) show that every number $\geqslant 88$ is good. (*a*) Suppose 87 were good. Then we would have $87 = 5x + 23y$ for some $x, y \geqslant 0$. But then $y \leqslant 3$ (since $87 < 23 \times 4$). If $y = 0$, then $x = 87/5$; if $y = 1$, then $x = 64/5$; if $y = 2$, then $x = 41/5$; if $y = 3$, then $x = 18/5$. In none of these cases is x a whole number. Therefore 87 cannot in fact be good and so must be bad. (*b*) To show that every number $\geqslant 88$ is good you only need to check that $88 = 1 \times 23 + 13 \times 5$, $89 = 3 \times 23 + 4 \times 5$, $90 = 0 \times 23 + 18 \times 5$, $91 = 2 \times 23 + 9 \times 5$, and $92 = 4 \times 23 + 0 \times 5$ are all good.

Exercise 15   (i) Suppose $4b - 5 = 79$ were good. Then $79 = 5x + 21y$ for some $x, y \geqslant 0$. But then $y \leqslant 3$. If $y = 0$, $x = 79/5$; if $y = 1$, $x = 58/5$; if $y = 2$, $x = 37/5$; if $y = 3$, $x = 16/5$. In none of these cases is x a whole number, so 79 must in fact be bad.

(ii) Suppose $4b - 5 = 799$ were good. Then $799 = 5x + 201y$ for some $x, y \geqslant 0$. But then $y \leqslant 3$. If $y = 0$, $x = 799/5$; if $y = 1$, $x = 598/5$; if $y = 2$, $x = 397/5$; if $y = 3$, $x = 196/5$. In none of these cases is x a whole number, so 799 must in fact be bad.

(iii) Suppose $4b - 5 = 7999$ were good. Then $7999 = 5x + 2001y$ for some $x, y \geqslant 0$. But then $y \leqslant 3$ and x cannot be a whole number. Hence 7999 must in fact be bad.

Exercise 16   Suppose $4b - 5$ were good. Then $4b - 5 = 5x + by$ for some

$x, y \geq 0$. But then $y \leq 3$ (because $4b - 5 < b \times 4$). If $y = 0$, $x = (4b - 5)/5$; if $y = 1$, $x = (3b - 5)/5$; if $y = 2$, $x = (2b - 5)/5$; if $y = 3$, $x = (b - 5)/5$. In none of these cases is x a whole number. (Why not?) So $4b - 5$ must in fact be bad.

Exercise 17 (i) (a) $4b - 4 = 4(5q + 1) - 4 = 20q = 4q \times 5 + 0 \times b$.
(b) $4b - 3 = 4(5q + 1) - 3 = 20q + 1 = 3q \times 5 + 1 \times b$.
(c) $4b - 2 = 4(5q + 1) - 2 = 20q + 2 = 2q \times 5 + 2 \times b$.
(d) $4b - 1 = 4(5q + 1) - 1 = 20q + 3 = q \times 5 + 3 \times b$.
 (ii) (a) $4b - 3 = 4(5q + 2) - 3 = 20q + 5 = (4q + 1) \times 5 + 0 \times b$.
(b) $4b - 1 = 4(5q + 2) - 1 = 20q + 7 = (3q + 1) \times 5 + 1 \times b$.
(c) $4b - 4 = 4(5q + 2) - 4 = 20q + 4 = 2q \times 5 + 2 \times b$.
(d) $4b - 2 = 4(5q + 2) - 2 = 20q + 6 = q \times 5 + 3 \times b$.
 (iii) (a) $4b - 2 = 4(5q + 3) - 2 = 20q + 10 = (4q + 2) \times 5 + 0 \times b$
(b) $4b - 4 = 4(5q + 3) - 4 = 20q + 8 = (4q + 1) \times 5 + 1 \times b$.
(c) $4b - 1 = 4(5q + 3) - 1 = 20q + 11 = (2q + 1) \times 5 + 2 \times b$.
(d) $4b - 3 = 4(5q + 3) - 3 = 20q + 9 = q \times 5 + 3 \times b$.
 (iv) (a) $4b - 1 = 4(5q + 4) - 1 = 20q + 15 = (4q + 3) \times 5 + 0 \times b$.
(b) $4b - 2 = 4(5q + 4) - 2 = 20q + 14 = (3q + 2) \times 5 + 1 \times b$.
(c) $4b - 3 = 4(5q + 4) - 3 = 20q + 13 = (2q + 1) \times 5 + 2 \times b$,
(d) $4b - 4 = 4(5q + 4) - 4 = 20q + 12 = q \times 5 + 3 \times b$.

Exercise 18 (i) If $a = 6$ and $b = 2n$ is even, then every good number $xa + yb$ is even. But then all odd numbers are bad, so the bad numbers never come to an end.

 (ii) If $a = 6$ and $b = 3n$ is a multiple of 3, then every good number is a multiple of 3. But then all numbers which are not multiples of 3 are bad, so the bad numbers never come to an end.

Exercise 19 (i) Suppose $5b - 6 = 6x + by$. Then $y \leq 4$. If $y = 0$, $x = (5b - 6)/6$; if $y = 1$, $x = (4b - 6)/6$; if $y = 2$, $x = (3b - 6)/6$; if $y = 3$, $x = (2b - 6)/6$; if $y = 4$, $x = (b - 6)/6$. In none of these cases is x a whole number, so $5b - 6$ must in fact be bad.

 (ii) (a) $b = 6q + r$ for some $r \leq 5$. If $r = 0$, 2, or 4, then b would be a multiple of 2; if $r = 3$, then b would be a multiple of 3. Hence $r = 1$ or $r = 5$.
(b) Suppose $r = 1$. Then $5b - 5 = 5(6q + 1) - 5 = 30q = 5q \times 6 + 0 \times b$; $5b - 4 = 5(6q + 1) - 4 = 30q + 1 = 4q \times 6 + 1 \times b$; $5b - 3 = 5(6q + 1) - 3 = 30q + 2 = 3q \times 6 + 2 \times b$; $5b - 2 = 5(6q + 1) - 2 = 30q + 3 = 2q \times 6 + 3 \times b$; $5b - 1 = 5(6q + 1) - 1 = 30q + 4 = q \times 6 + 4 \times b$. (c) Suppose $r = 5$. Then $5b - 5 = 5(6q + 5) - 5 = 30q + 20 = q \times 6 + 4 \times b$; $5b - 4 = 5(6q + 5) - 4 = 30q + 21 = (2q + 1) \times 6 + 3 \times b$; $5b - 3 = 5(6q + 5) - 3 = 30q + 22 = (3q + 2) \times 6 + 2 \times b$; $5b - 2 = 5(6q + 5) - 2 = 30q + 23 = (4q + 3) \times 6 + 1 \times b$; $5b - 1 = 5(6q + 5) - 1 = 30q + 24 = (5q + 4) \times 6 + 0 \times b$.

 (iii) $5b - 6$ is always bad by (i), and the six numbers $5b - 5$, $5b - 4$, $5b - 3$, $5b - 2$, $5b - 1$, $5b$ are always good by (ii), so all numbers $\geq 5b - 5$ are good.

Project (i) Suppose $ab - (a + b) = ax + by$ $(x, y \geq 0)$. Then $ab - (a + b) < (a - 1)b$, so $y \leq a - 2$. Thus $0 < b(a - 1 - y) = a(x - 1)$. But then a and b must have a common factor (a cannot divide $a - 1 - y$ since $1 \leq a - 1 - y \leq a - 1$). Hence $ab - (a + b)$ must in fact be bad.

(ii) (*a*) If *a* and *r* are both multiples of *c*, then so is $b = qa + r$, contradicting the assumption that *a* and *b* have no common factors. (*b*) Note that none of the numbers $r, 2r, 3r, \ldots, (a-1)r$ is divisible by *a*, and that no two of these numbers leave the same remainder when divided by *a* (if $i < j$ and *ir*, *jr* left the same remainder, then $(j-i)r$ would be a multiple of *a*). But there are only $a-1$ possible remainders, so each remainder occurs exactly once in the list $r, 2r, 3r, \ldots, (a-1)r$. So if $1 \le k \le a-1$, then some *mr* $(1 \le m \le a-1)$ leaves remainder exactly $k: mr = na + k$. But then $(a-1)b - k$ is good, because $(a-1)b - k = (a-1)(qa+r) - k = (a-1)qa + (a-m-1)r + mr - k = (a-m-1)(qa+r) + (mq+n)a = (a-m-1)b + (mq+n)a$. Hence $(a-1)b, (a-1)b-1, (a-1)b-2, \ldots, (a-1)b-(a-1)$ are all good.

(iii) $ab - (a+b)$ is always bad by (i), and the next *a* numbers are all good by (ii).

14 One picture is worth a thousand words

As a Geometrical Problem is handled Arithmetically
(by Algebraists) so an Arithmetical Question may
be handled Geometrically.
(David Gregory, *Memorandum*, 1706)

In the Postage Stamp Problem you have to combine given numbers (such as $a = 5$ and $b = 8$) to obtain certain other numbers (such as 26). On the face of it this looks like a simple problem in *arithmetic*. So in your first attempt to come to grips with the problem in Chapters 12 and 13 you took it more or less for granted that you should use the language and methods of *arithmetic*. And when you went beyond mere arithmetic and allowed first b, and then both a and b, to be unspecified numbers, you used *generalized arithmetic*—which is just another name for algebra. This simple-minded approach worked fairly well at first (in Exercises 1–19). But even those of you who had a really good go at the Project at the end of Chapter 13 probably felt that the algebra was getting a bit out of hand.

The challenge in this chapter is exactly the same as in the previous chapter: *to explain why there always seems to be a cut-off point beyond which all numbers are good* (provided only that a and b have no common factors). However in this chapter we want to try to find some 'obvious' explanation for this phenomenon, if necessary by looking at the problem in an entirely new way. In one of Martin Gardner's books on mathematical puzzles (*Aha! Insight*; published by 'Scientific American' in 1978) he tries to show how a problem which looks rather difficult can sometimes become very easy if only one finds the right way of looking at it. If possible we would like to find just such an 'Aha!' solution to the Postage Stamp Problem.

Finding 'the right way of *looking at* a problem' is often another way of saying that one has found a simple way of representing the problem *pictorially*. But if you have just slogged your way through Chapters 12 and 13, then you will not be surprised to hear that, though this sounds like getting something for nothing, there is in fact a catch! For though everything may indeed look obvious once the correct viewpoint has been discovered, *it is usually quite a struggle to find the right viewpoint in the first place.* As an indication of just how hard it can be to find the right viewpoint have a go at the next two exercises, and try them on your friends.

Exercise 20* What does Fig. 14.1 represent?

Fig. 14.1 After P. B. Porter in *The American Journal of Psychology,* Vol. 67, p. 550, 1954.

Exercise 21* What does Fig. 14.2 represent?

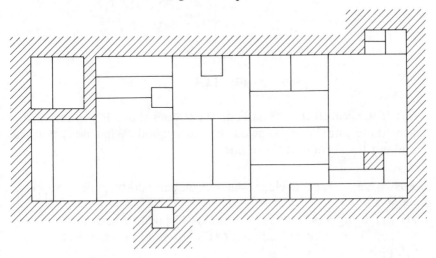

Fig. 14.2

The next two exercises introduce the 'square dot lattice'. You will need a supply of square dotty paper as in Fig. 14.3

Fig. 14.3

Exercise 22* Take a sheet of square dotty paper. Suppose $a = 3$ and $b = 7$.

(i) Show that 11 is bad and 16 is good.

(ii) Put axes on your sheet of square dotty paper (Fig. 14.4) and plot the lines $3x + 7y = 11$ and $3x + 7y = 16$.

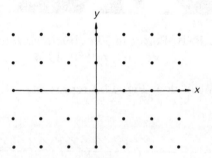

Fig. 14.4

(iii) You showed that 11 is bad. What does this tell you about the first of these lines? You showed that 16 is good. What does this tell you about the second of these lines?

Exercises 23* Take a clean sheet of square dotty paper. Suppose $a = 4$ and $b = 7$.

(i) (*a*) Is 16 good or bad?   (*b*) Is 17 good or bad?
    (*c*) Is 18 good or bad?   (*d*) Is 19 good or bad?

(ii) Plot each of the lines
    (*a*) $4x + 7y = 16$ (*b*) $4x + 7y = 17$
    (*c*) $4x + 7y = 18$ (*d*) $4x + 7y = 19$

on your sheet of dotty paper.

(iii) Interpret each of your answers to Part (i) as a statement about the corresponding line in Part (ii).

The point of view we shall adopt exploits the relationship between the following geometrical ideas.

(1) The first geometrical idea you must fix in your mind is *the set of points in the plane whose coordinates x, y are both whole numbers* (Fig. 14.5(i)). We shall call these points **lattice points**. We shall be particularly interested in *lattice points (x, y) with both coordinates x, y ⩾ 0*, and we shall refer to these as **lattice points in the positive quadrant.**

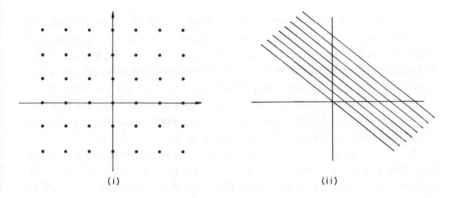

(i)   (ii)

Fig. 14.5

(2) The second geometrical idea you must fix in your mind is *the family of parallel lines ax + by = c*, where *a* and *b* are fixed positive whole numbers and *c* may be any whole number ⩾0 (Fig. 14.5(ii)).

Combining these two geometrical ideas we see that 'c is good' means simply that 'the line $ax + by = c$ passes through some lattice point (x, y) in the positive quadrant'. The next two exercises show how this geometrical interpretation works for specific values of *a* and *b*.

Exercise 24 Let $a = 5$ and $b = 2$. Draw a single pair of axes on a sheet of square dotty paper. Then draw the five lines, $5x + 2y = 0$, $5x + 2y = 1$, $5x + 2y = 2$, $5x + 2y = 3$, and $5x + 2y = 4$, on this piece of paper.

(i) Which of these five lines goes through one or more lattice points in the positive quadrant?

(ii) Which of the five numbers 0, 1, 2, 3, and 4 are good?

Exercise 25 Let $a = 5$ and $b = 3$. Draw a single pair of axes on a sheet of square dotty paper. Then draw the ten lines $5x + 3y = 0$, $5x + 3y = 1$, $5x + 3y = 2$, $5x + 3y = 3$, $5x + 3y = 4$, $5x + 3y = 5$, $5x + 3y = 6$, $5x + 3y = 7$, $5x + 3y = 8$, and $5x + 3y = 9$ on this piece of paper.

(i) Which of these ten lines goes through one or more lattice points in the positive quadrant?

(ii) Which of the ten numbers, 0, 1, 2, 3, 4, 5, 6, 7, 8, and 9, are good?

Your task in this chapter is to find an 'Aha!' reason why all numbers beyond a certain point are good. In the first instance we shall not worry too much about the *exact value* of the cut-off point, but will be content to show how our geometrical viewpoint provides a very simple 'Aha!' reason why there must in fact always be a cut-off point. We shall do this by showing in this chapter that *all numbers* $\geqslant ab$ *are good*. In the next chapter we shall come back to the question of finding the exact value of the cut-off point.

To show that all numbers $\geqslant ab$ are good it would be enough to show that the a numbers ab, $ab + 1, \ldots, ab + (a - 1)$, are all good. We shall in fact do much better than this and shall show that each of the ab numbers

$$ab, ab + 1, ab + 2, \ldots, 2ab - 1$$

is good. And how will we do this? By showing that each of the lines $ax + by = ab$, $ax + by = ab + 1$, $ax + by = ab + 2, \ldots, ax + by = 2ab - 1$ goes through *at least one* lattice point in the positive quadrant.

But before we can show this we must first answer one simple question about lattice points. Suppose the line $ax + by = c$ goes through more than one lattice point. *How close together can the lattice points on this line actually be?* Let us for the time being forget that we are mainly interested in 'lattice points in the positive quadrant'—for it will be easier to answer this question if we think about arbitrary lattice points (x, y) in the plane where x, y may be any whole numbers (positive, negative, or zero).

Exercise 26* Let $a = 5$ and $b = 2$.

(i) (a) Find all the lattice points (x, y) which lie on the line $5x + 2y = 0$.

(*b*) As you travel along the line $5x + 2y = 0$, what is the distance between each lattice point on the line and the next one you come to?

(ii) (*a*) Find all the lattice points (x, y) which lie on the line $5x + 2y = 10$.

(*b*) As you travel along the line $5x + 2y = 10$, what is the distance between each lattice point on the line and the next one you come to?

(iii) (*a*) Find all the lattice points (x, y) which lie on the line $5x + 2y = 11$.

(*b*) As you travel along the line $5x + 2y = 11$, what is the distance between each lattice point on the line and the next one you come to?

Exercise 27* Let $a = 5$ and $b = 3$.

(i) (*a*) Find all the lattice points (x, y) which lie on the line $5x + 3y = 0$.

(*b*) As you travel along the line $5x + 3y = 0$, what is the distance between each lattice point on the line and the next one you come to?

(ii) (*a*) Find all the lattice points (x, y) which lie on the line $5x + 3y = 15$.

(*b*) As you travel along the line $5x + 3y = 15$, what is the distance between each lattice point on the line and the next one you come to?

(iii) (*a*) Find all the lattice points (x, y) which lie on the line $5x + 3y = 14$.

(*b*) As you travel along the line $5x + 3y = 14$, what is the distance between each lattice point on the line and the next one you come to?

You should now be in a position to answer the question about the distance between lattice points on the line $ax + by = c$. The very simple answer which you will find in Exercise 29 is all you need to discover the 'Aha!' reason why all numbers $\geqslant ab$ are good (Exercise 32).

Exercise 28 Let a and b be positive whole numbers *with no common factors*.

(i) (*a*) Find all the lattice points (x, y) which lie on the line $ax + by = 0$.

(*b*) As you travel along the line $ax + by = 0$, what is the distance between each lattice point and the next one?

(ii) (*a*) Find all the lattice points (x, y) which lie on the line $ax + by = ab$.

(b) As you travel along the line $ax + by = ab$, what is the distance between each lattice point and the next?

Exercise 29* Let a and b be positive whole numbers with no common factors. Show that the distance between two lattice points on the line $ax + by = c$ must be $\geqslant \sqrt{(a^2 + b^2)}$.

We shall now show how this simple fact leads to the 'Aha!' solution you are looking for. You will find it helpful to work through one or two examples before looking at the general case.

Exercise 30 Let $a = 5$ and $b = 4$.

(i) How many lattice points are there in the rectangle with corners $(0, 0)$, $(4, 0)$, $(4, 5)$, and $(0, 5)$ if we *include* the bottom and left hand edges, but *exclude* the top and right hand edges (Fig. 14.6(i))? (In particular the corners $(0, 5)$ and $(4, 0)$ are excluded.)

(ii) How many lattice points are there in the parallelogram with corners $(4, 0)$, $(8, 0)$, $(4, 5)$, and $(0, 5)$ if we *include* the bottom and left hand edges but exclude the top and right-hand edges (Fig. 14.6(ii))?

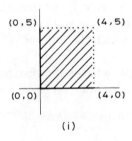 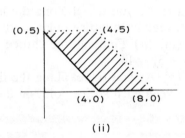

Fig. 14.6

(iii) For which whole numbers c does the line $5x + 4y = c$ intersect this parallelogram? (Remember that the right-hand edge is missing!)

(iv) Let $5x + 4y = c$ be a line which intersects the parallelogram in Fig. 14.6(ii). How many lattice points can lie both in the parallelogram and on this line? (Remember that the top edge of the parallelogram is missing!)

(v) By comparing the number of lattice points in the parallelogram, and the number of lines $5x + 4y = c$ which intersect the parallelogram, show that the numbers $20, 21, 22, \ldots, 39$ are all good.

Exercise 31 Let $a = 12$ and $b = 37$.

(i) How many lattice points are there in the rectangle with corners $(0, 0)$, $(37, 0)$, $(37, 12)$, and $(0, 12)$ if we *include* the bottom and left hand edges but *exclude* the top and right-hand edges (Fig. 14.7(i))?

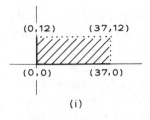

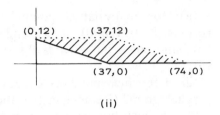

(i)   (ii)

Fig. 14.7

(ii) How many lattice points are there in the parallelogram with corners $(37, 0)$, $(74, 0)$, $(37, 12)$, and $(0, 12)$ if we *include* the bottom and left-hand edges but *exclude* the top and right-hand edges (Fig. 14.7(ii))?

(iii) For which whole numbers c does the line $12x + 37y = c$ intersect this parallelogram?

(iv) Let $12x + 37y = c$ be a line which intersects this paralleogram. How many lattice points can there be which are both in the parallelogram and also on this line?

(v) Show that the numbers $444 = 12 \times 37$, $12 \times 37 + 1$, $12 \times 37 + 2$, ..., $2 \times (12 \times 37) - 1 = 887$ are all good.

There is clearly nothing special about the pairs $a = 5$, $b = 4$ and $a = 12$, $b = 37$ in Exercises 30 and 31. The next exercise considers the general case.

Exercise 32* Let a and b be any two positive whole numbers with no common factors.

(i) How many lattice points are there in the rectangle with corners $(0, 0)$, $(b, 0)$, (b, a), and $(0, a)$ if we *include* the bottom and left-hand edges but *exclude* the top and right-hand edges (Fig. 14.8(i))?

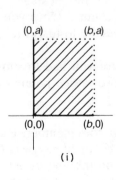

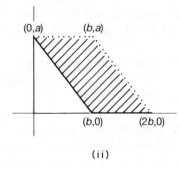

(i)   (ii)

Fig. 14.8

(ii) How many lattice points are there in the parallelogram with corners $(b, 0)$, $(2b, 0)$, (b, a), and $(0, a)$ if we *include* the bottom and left hand edges but *exclude* the top and right hand edges (Fig. 14.8 (ii))?

(iii) For which numbers c does the line $ax + by = c$ intersect this parallelogram? (Remember that the right hand edge is missing!) Show that each such line goes through at most one of the lattice points in the parallelogram.

(iv) By comparing the number of lattice points in the parallelogram and the number of lines $ax + by = c$ which intersect the parallelogram, show that the numbers ab, $ab + 1$, $ab + 2$, ... , $2ab - 1$ are all good.

It is worth going through the solution again to see whether it can honestly be called an 'Aha!' solution. Suppose you wanted to memorize the solution. What exactly does it involve? How many new tricks would you have to remember now that you know the right viewpoint? Let's try to list the various steps you have had to take in this chapter.

Step 1 First of all you had to realize that the statement 'c is good' has precisely the same meaning as 'the line $ax + by = c$ passes through *some lattice point* (x, y) in the positive quadrant'.

You already know that an algebraic equation like $ax + by = c$ can be thought of as the equation of a straight line. So the only new idea here is the idea of looking at *lattice points*—that is, points (x, y) both of whose coordinates are whole numbers. But even if this idea is new to you, it is only another way of saying that the Postage Stamp Problem is concerned with whole numbers of stamps. Once you have got used to this idea it should be almost impossible to forget it.

Step 2 When a and b have no common factor you had to show (in Exercise 29) that the distance between any two lattice points on the line $ax + by = c$ is $\geqslant \sqrt{(a^2 + b^2)}$.

Now this is something you will certainly have to remember. But it is a much simpler fact than the words themselves might suggest. For it simply says that 'if (x, y) is one lattice point on the line $ax + by = c$, then the next lattice point to the right is $(x + b, y - a)$ and the next

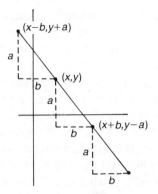

Fig. 14.9

lattice point to the left is $(x - b, y + a)$, and so on (as in Fig. 14.9).'
(When a and b have no common factors, then it is in fact true that
every line $ax + by = c$ goes through infinitely many different lattice
points. But we have at present no way of knowing this fact. So it may
be worth stressing that as far as Step 1 and Step 2 are concerned,
there may be lots of values of c of c for which the line $ax + by = c$
goes through *no lattice points at all*.)

You honestly don't need to know anything else. Once you have
understood Step 1 and Step 2 all you have to do is

(i) stare at a mental picture of the parallelogram which has its
corners at the points $(b, 0)$, $(2b, 0)$, (b, a), and $(0, a)$, and which has
its top and right-hand edges missing (Fig. 14.10);

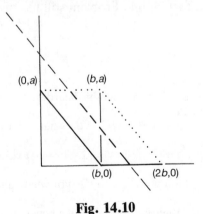

Fig. 14.10

(ii) 'see' that this parallelogram contains exactly ab lattice points;
(iii) realize that the line $ax + by = ab$ forms the left-hand edge of
this parallelogram, while the line $ax + by = 2ab$ forms the missing

right-hand edge, so there are exactly ab lines $ax + by = c$ (with c a whole number) which actually intersect the parallelogram, namely the lines $ax + by = ab$, $ax + by = ab + 1$, $ax + by = ab + 2, \ldots, ax + by = 2ab - 1$; and finally

(iv) observe that each of these lines intersects the parallelogram in a line segment which is *so short* that it can go through *at most one* of the ab lattice points in the parallelogram.

For if each of these ab lines goes through at most one of the ab lattice points in the parallelogram, then the only way that all ab of these lattice points can be used up is if each of the lines $ax + by = ab$, $ax + by = ab + 1$, $ax + by = ab + 2, \ldots, ax + by = 2ab - 1$ passes through exactly one of them. Since these lattice points all lie in the positive quadrant you then know that each of the numbers ab, $ab + 1$, $ab + 2, \ldots, 2ab - 1$ is good. But then all numbers $\geq ab$ are good.

There is next to nothing here for you to memorize. As long as you can remember

(1) that you are going to show that the numbers ab, $ab + 1$, $ab + 2, \ldots, 2ab - 1$ are all good, and

(2) that this is done by counting lattice points in some parallelogram,

then you are almost bound to draw the right parallelogram—and after that you can't really go wrong.

Anyway you can now be absolutely certain that there always is a last bad number, and that it is always $\leq ab - 1$. In the next chapter we shall finally pin down this last bad number exactly. We shall also discover that the Postage Stamp Problem still has one or two surprises in store.

Hints to exercises

20. Can you see his eyes? Can you see his beard?
21. (See after the hint to Exercise 32.)
26. (i) If $5x + 2y = 0$, then $y = -5x/2$. For what values of x is y a whole number?

   (ii) If $5x + 2y = 10$, then $y = (10 - 5x)/2$. For what values of x is y a whole number?

   (iii) If $5x + 2y = 11$, then $y = (11 - 5x)/2$. For what values of x is y a whole number?
27. (iii) If $5x + 3y = 14$, then $y = (14 - 5x)/3$. Show that y is a whole number precisely when x is 1 more than a multiple of 3: say $x = 3m + 1$.
28. (i) If $ax + by = 0$, then $y = -ax/b$. For what values of x is y a whole number? (Remember that a and b have no common factors.) Sketch the line $ax + by = 0$ and mark all the lattice points on it.

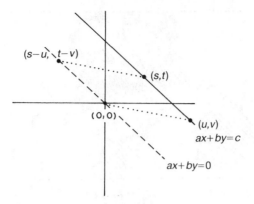

Fig. 14.11

(ii) If $ax + by = ab$, then $y = (ab - ax)/b$. For what values of x is y a whole number?

29. Suppose (s, t) and (u, v) are two lattice points on the line $ax + by = c$ (Fig. 14.11).

(i) Show that $(s - u, t - v)$ is a lattice point on the line $ax + by = 0$.

(ii) Show that the distance between the two lattice points (s, t) and (u, v) on the line $ax + by = c$ is equal to the distance between the two lattice points $(s - u, t - v)$ and $(0, 0)$ on the line $ax + by = 0$.

30. (ii) Cut the rectangle in Part (i) into two pieces along a diagonal, and reassemble them to make the required parallelogram.

(iii) The line $5x + 4y = 20$ is one of the edges included in the parallelogram, and the line $5x + 4y = 40$ is one of the edges excluded from the parallelogram.

(iv) Use Exercise 29.

(v) There are *twenty* lattice points and *twenty* lines, and each line passes through *at most one* lattice point. So . . . .

31. Imitate your solution to Exercise 30.

32. Imitate your solution to Exercise 30.

21. It's a map of Europe. Now try to identify the countries.

Solutions

Exercise 20 See Fig. 14.12

Exercise 21 See Fig. 14.14.

Exercise 22 (i) Suppose $11 = 3x + 7y$. Then $y \leq 1$. If $y = 0$, $x = 11/3$; if $y = 1$, $x = 4/3$. In neither case is x a whole number. So 11 must be bad. $16 = 3 \times 3 + 1 \times 7$ is good.

(iii) The line $3x + 7y = 11$ does not pass through any dots in the positive quadrant. The line $3x + 7y = 16$ passes through at least one dot in the positive quadrant (e.g. $(3, 1)$).

Fig. 14.12

Exercise 23 (i) (*a*) Good. (*b*) Bad. (*c*) Good. (*d*) Good.

(iii) (*a*) The line $4x + 7y = 16$ passes through at least one dot in the positive quadrant, namely $(4, 0)$. (*b*) The line $4x + 7y = 17$ does not pass through any dots in the positive quadrant. (*c*) The line $4x + 7y = 18$ passes through at least one dot in the positive quadrant, namely $(2, 1)$. (*d*) The line $4x + 7y = 19$ passes through at least one dot in the positive quadrant, namely $(3, 1)$.

Exercise 24 (i) The lines $5x + 2y = 0$, $5x + 2y = 2$, and $5x + 2y = 4$ pass through at least one lattice point in the positive quadrant; the lines $5x + 2y = 1$ and $5x + 2y = 3$ pass through no lattice points in the positive quadrant.

(ii) 0, 2, 4 are good; 1, 3 are bad.

Exercise 25 (i) The lines $5x + 3y = 0$, $5x + 3y = 3$, $5x + 3y = 5$, $5x + 3y = 6$, $5x + 3y = 8$, $5x + 3y = 9$ all pass through at least one lattice point in the positive quadrant; the lines $5x + 3y = 1$, $5x + 3y = 2$, $5x + 3y = 4$, $5x + 3y = 7$ pass through no lattice points in the positive quadrant.

(ii) 0, 3, 5, 6, 8, 9 are good; 1, 2, 4, 7 are bad.

Exercise 26 (i) (*a*) If $5x + 2y = 0$, then $y = -5x/2$. For y to be a whole number, x must be even: so $x = 2m$, $y = -5m$. Hence the lattice points on the line $5x + 2y = 0$ are precisely the points $(2m, -5m)$, where m may be any whole number (positive, negative, or zero). (*b*) The distance between successive lattice points $(2m, -5m)$, $(2m + 2, -5m - 5)$ is therefore $\sqrt{(2^2 + 5^2)} = \sqrt{29}$.

(ii) (*a*) If $5x + 2y = 10$, then $y = (10 - 5x)/2$. For y to be a whole number, x must be even: so $x = 2m$, $y = 5 - 5m$. Hence the lattice points on the line $5x + 2y = 10$ are precisely the points $(2m, 5 - 5m)$, where m may be any whole number (positive, negative, or zero). (*b*) The distance between successive lattice points $(2m, 5 - 5m)$, $(2m + 2, -5m)$ is therefore $\sqrt{(2^2 + 5^2)} = \sqrt{29}$.

(iii) (*a*) If $5x + 2y = 11$, then $y = (11 - 5x)/2$. For y to be a whole number, x must be odd: so $x = 1 + 2m$, $y = 3 - 5m$. Hence the lattice points on the line $5x + 2y = 11$ are precisely the points $(1 + 2m, 3 - 5m)$, where m may be any whole number. (*b*) The distance between successive lattice points $(1 + 2m, 3 - 5m)$, $(3 + 2m, -2 - 5m)$ is therefore $\sqrt{(2^2 + 5^2)} = \sqrt{29}$.

Exercise 27 (i) (*a*) If $5x + 3y = 0$, then $y = -5x/3$. For y to be a whole number, x must be a multiple of 3: so $x = 3m$, $y = -5m$. Hence the lattice points on the line $5x + 3y = 0$ are precisely the points $(3m, -5m)$, where m may be any whole number. (*b*) The distance between successive lattice points $(3m, -5m)$, $(3m + 3, -5m - 5)$ is $\sqrt{(3^2 + 5^2)} = \sqrt{34}$.

(ii) (*a*) If $5x + 3y = 15$, then $y = (15 - 5x)/3$. Hence the lattice points on the line $5x + 3y = 15$ are precisely the points $(3m, 5 - 5m)$, where m may be any whole number. (*b*) The distance between successive lattice points $(3m, 5 - 5m)$, $(3m + 3, -5m)$ is $\sqrt{(3^2 + 5^2)} = \sqrt{34}$.

(iii) (*a*) If $5x + 3y = 14$, then $y = (14 - 5x)/3$. Hence the lattice points on the line $5x + 3y = 14$ are precisely the points $(1 + 3m, 3 - 5m)$, where m may be any whole number. (*b*) The distance between successive lattice points $(1 + 3m, 3 - 5m)$, $(4 + 3m, -2 - 5m)$ is $\sqrt{(3^2 + 5^2)} = \sqrt{34}$.

Exercise 28 (i) (*a*) If $ax + by = 0$, then $y = -ax/b$. If y is to be a whole number, then x must be a multiple of b (because a and b have no common factors): so $x = bm$, $y = -am$, and the lattice points on the line $ax + by = 0$ are precisely the points $(bm, -am)$, where m may be any whole number. (*b*) The distance between successive lattice points $(bm, -am)$, $(b(m + 1), -a(m + 1))$ is $\sqrt{(b^2 + a^2)}$.

(ii) (*a*) If $ax + by = ab$, then $y = (ab - ax)/b$. For y to be a whole number, x must be a multiple of b: so $x = bm$, $y = a - am$, and the lattice points on the line $ax + by = ab$ are precisely the points $(bm, a - am)$, where m may be any whole number. (*b*) The distance between successive lattice points $(bm, a - am)$, $(b(m + 1), -am)$ is $\sqrt{(b^2 + a^2)}$.

Exercise 29 If (s, t), (u, v) are two lattice points on the line $ax + by = c$, then $as + bt = c$ and $au + bv = c$, so $a(s - u) + b(t - v) = 0$. But then $(s - u, t - v)$ is a lattice point on the line $ax + by = 0$. The four points, (s, t), (u, v), $(0, 0)$ and $(s - u, t - v)$, form the vertices of a parallelogram, so the distance between (s, t) and (u, v) is equal to the distance between $(s - u, t - v)$ and $(0, 0)$. Since $(s - u, t - v)$ and $(0, 0)$ are both lattice points on the line $ax + by = 0$, you already know from Exercise 28(i) that the distance between them is $\geqslant \sqrt{(b^2 + a^2)}$.

Exercise 30 (i) 5 rows, 4 lattice points in each row, so $5 \times 4 = 20$ lattice points altogether.

(ii) Exactly the same number as in the rectangle with corners $(0, 0)$, $(4, 0)$, $(4, 5)$, and $(0, 5)$ (see Fig. 14.13(i)), namely $5 \times 4 = 20$.

(iii) Precisely when $c = 20, 21, 22, \ldots, 39$.

(iv) Each line which intersects the parallelogram does so in a line segment which has length $\sqrt{(5^2 + 4^2)}$ and which has its lower endpoint missing. Exercise 29 then shows that each such line segment passes through *at most one* lattice point.

(v) On the one hand the parallelogram contains exactly $5 \times 4 = 20$ lattice points (x, y); and each of these lattice points lies on some line $5x + 4y = c$ with c a whole number. On the other hand there are exactly 20 whole numbers c for which the line $5x + 4y = c$ intersects the parallelogram; and

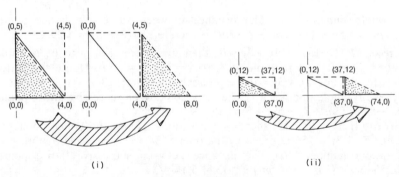

Fig. 14.13

each of these lines passes through *at most one* of the 20 lattice points in the parallelogram. The only way both of these facts can be true is if each of the lines $5x + 4y = 20$, $5x + 4y = 21$, ..., $5x + 4y = 39$ passes through *exactly one* lattice point in the parallelogram. Since all these lattice points lie in the positive quadrant, the numbers 20, 21, 22, ..., 39 are all good.

Exercise 31 (i) 12 rows, 37 lattice points in each row, so 12×37 lattice points altogether.

(ii) Exactly the same number as there are in the rectangle with corners $(0, 0)$, $(37, 0)$, $(37, 12)$, and $(0, 12)$ (see Fig. 14.13(ii)), namely $12 \times 37 = 444$.

(iii) Precisely when $c = 12 \times 37$, $12 \times 37 + 1$, $12 \times 37 + 2$, ..., $2(12 \times 37) - 1$.

(iv) Each line which intersects the parallelogram does so in a line segment which has length $\sqrt{(12^2 + 37^2)}$ and which has its lower endpoint missing. Exercise 29 then shows that each such line segment passes through *at most one* lattice point.

(v) On the one hand the parallelogram contains exactly $12 \times 37 = 444$ lattice points (x, y); and each of these lattice points lies on some line $12x + 37y = c$ with c a whole number. On the other hand there are exactly 12×37 whole numbers c for which the line $12x + 37y = c$ intersects the parallelogram; and each of these lines passes through *at most one* of the 12×37 lattice points in the parallelogram. The only way both of these facts can be true is if each of the lines $12x + 37y = 12 \times 37$, $12x + 37y = 12 \times 37 + 1$, ..., $12x + 37y = 2(12 \times 37) - 1$ passes through *exactly one* lattice point in the parallelogram. Since all these lattice points lie in the positive quadrant, the numbers 12×37, $12 \times 37 + 1$, ..., $2(12 \times 37) - 1$ are all good.

Exercise 32 (i) a rows, b lattice points in each row, so $a \times b$ lattice points altogether.

(ii) Exactly the same number as there are in the rectangle with corners $(0, 0)$, $(b, 0)$, (b, a), and $(0, a)$.

(iii) Precisely when $c = a \times b$, $a \times b + 1$, ..., $2(a \times b) - 1$. Each such line intersects the parallelogram in a line segment of length $\sqrt{(b^2 + a^2)}$ which has its lower endpoint missing. Exercise 29 then shows that each such line segment passes through *at most one* lattice point.

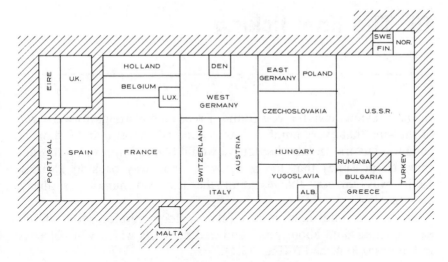

Fig. 14.14

(iv) The parallelogram contains exactly $a \times b$ lattice points (x, y), and each lies on some line $ax + by = c$ with c a whole number. But there are precisely $a \times b$ whole numbers c for which the line $ax + by = c$ intersects the parallelogram, and each of these lines passes through *at most one* of the $a \times b$ lattice points in the parallelogram. Hence each of the lines $ax + by = ab$, $ax + by = ab + 1, \ldots, ax + by = 2ab - 1$ passes through *exactly one* of the lattice points in the parallelogram, so the numbers $ab, ab + 1, \ldots, 2ab - 1$ are all good.

15 The final licking

In the previous chapter you found a reasonably straightforward way of showing that every number $\geq ab$ is good whenever a and b have no common factors. In this chapter we would like to improve on this and show (if possible) that the guess you made way back in Exercise 10(iii) was correct: that is, that the last bad number is always $ab - (a + b)$.

You need one new idea. Take a long careful look at the experimental data about good and bad numbers which you collected in Exercises 3, 4, and 5 (Fig. 15.1).

$a = 5, b = 2$

Number	0	1	2	3	4	5
Good?	✓		✓		✓	✓
Bad?		×		×		

$a = 5, b = 3$

Number	0	1	2	3	4	5	6	7	8	9	10
Good?	✓			✓		✓	✓		✓	✓	✓
Bad?		×	×		×			×			

$a = 5, b = 4$

Number	0	1	2	3	4	5	6	7	8	9	10	11	12	13	14	15
Good?	✓				✓	✓			✓	✓	✓		✓	✓	✓	✓
Bad?		×	×	×			×	×				×				

Fig. 15.1

Exercise 33* (i) You already know that as soon as you come across *b* ticks in a row you have passed the cut-off point. But can you see anything surprising about the distribution of good and bad numbers *up to the cut-off point*?

(ii) In Exercises 6, 7, and 8 you drew up a table like the ones in Fig. 15.1 for the following pairs:

$a = 5, b = 6$; $a = 5, b = 7$; $a = 5, b = 8$; $a = 5, b = 9$;

$a = 5, b = 10$; $a = 5, b = 11$; $a = 5, b = 12$; $a = 5, b = 13$.

Do the good and bad numbers between 0 and the cut-off point in these other tables show the same surprising pattern?

(iii) Suppose $a = 7$ and $b = 11$. Draw up a table like the one in Exercise 5. Do the good and bad numbers between 0 and the cut-off point show the same pattern?

In Exercises 3, 4, and 5 we began each table by testing the number 0. At the time you may have felt that this was a bit silly. After all 0 is always good, so it hardly seems to be worth writing it down each time. Would it have really mattered if you had left it out? Suppose, for example, that you had omitted the number 0 from the table you completed in Exercise 5. What difference would it have made (Fig. 15.2)? Perhaps not all that much.

$$a = 5, b = 4$$

Number		1	2	3	4	5	6	7	8	9	10	11	12	13	14	15
Good?					✓	✓			✓	✓	✓		✓	✓	✓	✓
Bad?		×	×	×			×	×				×				

Fig. 15.2

It would have been harder to spot the pattern in Exercise 33(i). But the pattern would still have been partly visible. And once you had noticed it you would have been more or less compelled to put the number 0 back in just to complete this symmetry of good and bad numbers. For it looks as though we should be trying to explain why the number 0, which is always the first good number, appears to match up with the number $ab - (a + b)$, which we suspect is always the last bad number.

Exercise 34 Suppose $a = 5$ and $b = 4$. Which number matches up with 11? Which number matches up with 12? Which number matches up with 13? . . .

You have not yet proved that the good numbers and bad numbers between 0 and the cut-off point always occur in this symmetrical way. But the idea fits in rather nicely with what you *have* already proved. For you showed in Chapter 14 that there always is a last bad number, and that it is always $<ab$: let's call this last bad number B. Then B is *bad* and every number $>B$ is *good*. On the other hand 0 is obviously *good* and every number <0 is *bad*. So you have already proved this symmetry between good and bad numbers for numbers ≤ 0 and numbers $\geq B$ (Fig. 15.3). This fits in very nicely with your guess about the symmetry of good and bad numbers which lie between 0 and B.

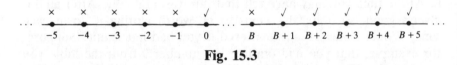

Fig. 15.3

But wait a moment! You only wanted to show that the last bad number B is always equal to $ab - (a + b)$. Why introduce all this stuff about the symmetry between good and bad numbers? There is after all no obvious connection between the two things. But you certainly need some sort of new idea. And once you have noticed something as striking as this totally unexpected symmetry of good and bad numbers, it is hard to believe that it is completely irrelevant. Given time you might come up with an even better idea. But meanwhile it seems worth following up the one idea you have had.

What exactly would you like to prove? When a and b are positive whole numbers with no common factor, you want to show

(I) that the last bad number b is always equal to $ab - (a + b)$;

(II) that the good and bad numbers between 0 and $ab - (a + b)$ occur symmetrically in the sense that '*if c is good, then $ab - (a + b) - c$ is bad*', and '*if c is bad, then $ab - (a + b) - c$ is good*'.

Rather than attack these two problems head-on we shall begin by nibbling away at the bits we can do fairly easily. For example, you would like to show that '$ab - (a + b)$ is always bad'. Here are two ways of doing this.

Exercise 35 (i) *First method:* Show that the lattice points $(-1, a - 1)$ and $(b - 1, -1)$ both lie on the line $ax + by = ab - (a + b)$. Calculate the distance between these two points. Hence show that the line $ax + by = ab - (a + b)$ does not go through any lattice point in the positive quadrant. Deduce that $ab - (a + b)$ is bad.

(ii) *Second method:* Suppose it were possible to find whole numbers $x, y \geqslant 0$ for which $ab - (a + b) = xa + yb$. Show first that x, y would have to satisfy $x \leqslant b - 2$ and $y \leqslant a - 2$. Rearrange the equation $ab - (a + b) = xa + yb$ into the form

$$a(b - 1 - x) = b(y + 1).$$

Then use the fact that $0 < y + 1 < a$ to show that a and b would have to have a common factor. Deduce that $ab - (a + b)$ is bad whenever a and b have no common factor.

The same ideas can be used to deal with half of Problem (II) (the easy half!).

Exercise 36 (i) Let c be any number between 0 and $ab - (a + b)$. Show that 'if c is good, then $ab - (a + b) - c$ is bad'.

(ii) Deduce that at most half the numbers between 0 and $ab - (a + b)$ are good.

Exercise 37 Just suppose that you managed to prove that *exactly half* of the numbers between 0 and $ab - (a + b)$ are good. Show how you could use this to prove the other half of (II), namely that 'if c is bad, then $ab - (a + b) - c$ is good'.

So far you have only come across two ways of counting good numbers. The first is to write them all down and count them one by one. And the second is to think of good numbers in terms of lines which pass through lattice points in the positive quadrant and then to find some clever way of counting the lines and the lattice points. You used the lattice point idea in the previous chapter to show that all numbers $\geqslant ab$ are good. We shall now use it again to count the number of good numbers $\leqslant ab$. Let N stand for the number of good numbers $\leqslant ab$.

Exercise 38* You already know that *at most half* of the numbers c satisfying $0 \leqslant c \leqslant ab - (a + b)$ are good. You would like to prove two things at once simply by calculating the value of N:

(i) first that *exactly half* the numbers between 0 and $ab - (a + b)$ are good;

(ii) second that *all* numbers c satisfying $ab - (a + b) < c \leq ab$ are good.
What would you like the value of N to be?

The method we use to calculate N is surprisingly simple. Here is how it works when $a = 5$ and $b = 4$.

Exercise 39* Suppose $a = 5$ and $b = 4$.

(i) Each good number $c < ab = 20$ can be written in the form $c = 5x + 4y$ where $x, y \geq 0$. Show that each good number $c < ab = 20$ can be written like this in *exactly one way*.

(ii) Show that the good number $c = ab = 20$ can be written in the form $20 = 5x + 4y$ (with $x, y \geq 0$) in *exactly two different ways*.

(iii) The number 14 is good since the line $5x + 4y = 14$ passes through a lattice point in the positive quadrant—namely the lattice point $(2, 1)$. So instead of counting the good number 14 you could count the good line $5x + 4y = 14$ instead. But you showed in Part (i) that this good line passes through *exactly one* lattice point in the positive quadrant (namely, the point $(2, 1)$). So instead of counting the good line $5x + 4y = 14$ you could count the lattice point $(2, 1)$. This point is marked with a cross and labelled '14' in Fig. 15.4(i). Mark the other lattice points in the positive quadrant which represent good numbers <20, and label each one with the good number it represents.

(iv) Now mark the two lattice points which represent the good number $ab = 20$.

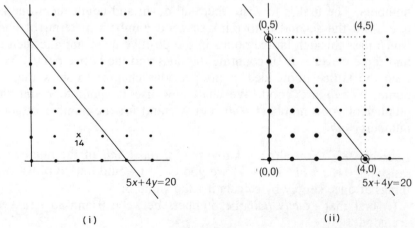

$5x+4y=20$

(i)

$5x+4y=20$

(ii)

Fig. 15.4

(v) Finally calculate N (that is, the number of good numbers $\leqslant 20$) by counting first *all* the lattice points which represent good numbers <20, and then *just one* of the two lattice points which represent the good number 20.

There would be no advantage in counting lattice points rather than good numbers themselves if you did not happen to notice the following simple fact: *the lattice points which represent good numbers $<ab = 20$, together with just one of the two lattice points which represent the good number $ab = 20$, use up exactly half of the lattice points in the rectangle which has its corners at (0, 0), (4, 0), (4, 5), and (0, 5) (Fig. 15.4(ii))*. Once you notice this, then the value of N can be calculated without counting each individual lattice point. For the number of lattice points in the whole rectangle is obviously $(4 + 1) \times (5 + 1) = 30$, so we must have

$$N = \tfrac{1}{2}(4 + 1) \times (5 + 1) = 15.$$

Exercise 40* Suppose $a = 5$ and $b = 6$.
(i) Show that each good number $c < ab = 30$ can be written in the form $c = 5x + 6y$ (where $x, y \geqslant 0$) in *exactly one way*.
(ii) Show that the good number $ab = 30$ can be written in the form $30 = 5x + 6y$ (where $x, y \geqslant 0$) in *exactly two different ways*.
(iii) Each good number $c < 30$ can now be represented by the unique lattice point (x, y) in the positive quadrant which lies on the line $5x + 6y = 30$. In Fig. 15.5 find the unique lattice point which

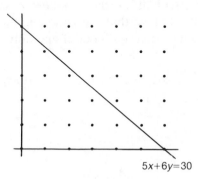

$5x+6y=30$

Fig. 15.5

represents the good number 17 and the two lattice points which represent the good number 30.
(iv) Find N.

Now that you have worked through Exercise 39 (with $a = 5$ and $b = 4$) and Exercise 40 (with $a = 5$ and $b = 6$) you should have no trouble at all with the general case.

Exercise 41* Suppose a and b are two positive whole numbers with no common factors.

(i) Show that each good number $c < ab$ can be written in the form $c = ax + by$ (where $x, y \geq 0$) in *exactly one way*.

(ii) Show that the good number $c = ab$ can be written in the form $ab = ax + by$ (where $x, y \geq 0$) in *exactly two different ways*.

(iii) Show that $N = \frac{1}{2} \times (a + 1) \times (b + 1)$.

In Exercise 38 you decided what value of N would allow you to prove at a single stroke

(i) that exactly half the numbers between 0 and $ab - (a + b)$ are good, and

(ii) that all numbers c satisfying $ab - (a + b) < c \leq ab$ are good.

In Exercise 41(iii) you then showed that N is equal to precisely this value. You should now be able to show

(I) that $ab - (a + b)$ is the last bad number, and

(II) that the good and bad numbers between 0 and $ab - (a + b)$ occur symmetrically.

Exercise 42* (i) Use the value $N = \frac{1}{2}(a + 1)(b + 1)$ to show that every number c satisfying $ab - (a + b) < c \leq ab$ is good. Deduce that $ab - (a + b)$ is the last bad number.

(ii) Show that exactly half of the numbers c satisfying $0 \leq c \leq ab - (a + b)$ are good. Deduce that the good and bad numbers occur symmetrically in the sense that 'c is good if and only if $ab - (a + b) - c$ is bad'.

Hints to exercises

33. (i) Read the tables between 0 and the cut-off point, first forwards, then backwards:

$a = 5, b = 2$ Forwards: good – bad – good – bad.
Backwards: bad – good – bad – good.

$a = 5, b = 3$ Forwards: good – bad – bad – good – bad – good – good – bad.
Backwards: bad – good – good – bad – good – bad – bad – good.

36. (i) *First method:* Rotate the parallelogram with corners $(0, 0)$, $(-1, a - 1)$, $(b - 1, -1)$, and $(b, -a)$ through $180°$ about its centre. This takes the line $ax + by = c$ to the line $ax + by = ab - (a + b) - c$. If the line $ax + by = c$ passes through the lattice point P in the positive quadrant, then the line $ax + by = ab - (a + b) - c$ must pass through the image of P under this half turn. Why does this imply that $ab - (a + b) - c$ is bad?

Second method: The algebra here is a bit messy but it's not too bad. If c is good then you can certainly write $c = xa + yb$ where x, y are whole numbers $\geqslant 0$. If you also know that $c \leqslant ab - (a + b)$, then x and y cannot be too big: in fact you must have $0 \leqslant x \leqslant$ ____, and $0 \leqslant y \leqslant$ ____. But $0 \leqslant ab - (a + b) - c \leqslant ab - (a + b)$. So if $ab - (a + b) - c$ were also good, then you could write $ab - (a + b) - c = ua + vb$ with $0 \leqslant u \leqslant$ ____, and $0 \leqslant v \leqslant$ ____ as before. Try to show that this would lead to an impossible conclusion (such as, 'a and b must have a common factor'). You can then be sure that $ab - (a + b) - c$ cannot be good.

37. The numbers between 0 and $ab - (a + b)$ come in pairs: c, $ab - (a + b) - c$. In Exercise 35 you showed that *at most one* number in each pair can be good.

39. (i) Suppose some number $c < ab = 20$ is good in two different ways: say $c = 5x + 4y$ and $c = 5u + 4v$. Then since $c < 20$ you can be sure that $0 \leqslant x, u \leqslant 3$ and $0 \leqslant y, v \leqslant 4$. Now try to show that this leads to an impossible conclusion unless $x = u$ and $y = v$.

(ii) You know that $c = 20$ is good in at least two ways: namely $20 = 5 \times 4 + 4 \times 0$ and $20 = 5 \times 0 + 4 \times 5$. Suppose 20 is good in some other way as well: say $20 = 5x + 4y$. Then $1 \leqslant x \leqslant 3$ and $1 \leqslant y \leqslant 4$. Show that this leads to an impossible conclusion.

40. Imitate your solution to Exercise 39.

41. Imitate your solution to Exercise 39.

42. (i) $\frac{1}{2}(a + 1)(b + 1) = N$
= (number of good numbers c satisfying $0 \leqslant c \leqslant ab$)
= (number of good numbers c satisfying $0 \leqslant c \leqslant ab - (a + b)$)
+ (number of good numbers c satisfying $ab - (a + b) < c \leqslant ab$)
$\leqslant \frac{1}{2}(ab - (a + b) + 1) + (ab - [ab - (a + b)]) = \ldots$.

Solutions

Exercise 33 (i) Between 0 and the cut-off point exactly half of the numbers are good and half are bad. More striking still is the fact that the *forward* sequence of 'goods' and 'bads' is exactly the same as the *reverse* sequence of 'bads' and 'goods'.

Exercise 34 11 matches up with 0, 12 matches up with -1, 13 matches up with -2, 14 matches up with -3, and so on. 12, 13, 14, ... are all good; $-1, -2, -3, \ldots$ are all bad.

Exercise 35 (i) Both $(-1, a - 1)$ and $(b - 1, -1)$ lie on the line $ax + by = ab - (a + b)$ and outside the positive quadrant, and the distance between them is $\sqrt{(b^2 + a^2)}$. So by Exercise 29 no lattice point can lie between them on the line $ax + by = ab - (a + b)$. Hence the line cannot pass through any lattice point in the positive quadrant, so the number $ab - (a + b)$ is bad.

(ii) If $ax + by = ab - (a + b)$, then $x < b - 1$; and rearranging the equation we get $a(b - 1 - x) = b(y + 1)$. But then b is a factor of the left-hand side. Since a and b have no common factors, b must divide $b - 1 - x$, which is impossible because $0 < b - 1 - x \leqslant b - 1$. Hence $ab - (a + b)$ cannot be good, and so must be bad.

Exercise 36 (i) We follow the hint. *First method:* Each line $ax + by = c$ which intersects the parallelogram with vertices $(0, 0)$, $(-1, a - 1)$, $(b - 1, -1)$, and $(b, -a)$ does so in a line segment of length $\sqrt{(b^2 + a^2)}$. By Exercise 29 each such line segment passes through either at most one lattice point, or exactly two lattice points—one at each endpoint of the segment. Let c be any number between 0 and $ab - (a + b)$, and suppose that P is a lattice point which lies both in the positive quadrant and on the line $ax + by = c$. The centre of the parallelogram is the point $((b - 1)/2, -\frac{1}{2})$, so the image of P under a $180°$ rotation about this centre is another lattice point and the image of the line $ax + by = c$ is the line $ax + by = ab - (a + b) - c$. If P lies in the positive quadrant, then the image of P does not. (Why not?) Hence the line $ax + by = ab - (a + b) - c$ passes through no lattice point in the positive quadrant. (Why not?) So $ab - (a + b) - c$ is bad. *Second method:* Suppose $c = xa + yb$ for some $x, y \geqslant 0$, and $ab - (a + b) - c = ua + vb$ for some $u, v \geqslant 0$. Then $ab - (a + b) - (xa + yb) = ua + vb$, so $a(b - (u + x + 1)) = b(v + y + 1)$. But then b must divide the left-hand side, so b must divide $(b - (u + x + 1))$ (since a and b have no common factors), which is impossible because $0 < b - (u + x + 1) < b$.

(ii) The numbers between 0 and $ab - (a + b)$ come in pairs: c, $ab - (a + b) - c$. At most one member of each pair is good. Therefore at most half of all the numbers between 0 and $ab - (a + b)$ are good.

Exercise 37 If exactly half the numbers between 0 and $ab - (a + b)$ were good, then exactly one member of each pair c, $ab - (a + b) - c$ would have to be good. So if c were bad, then $ab - (a + b) - c$ would have to be good.

Exercise 38 You expect all $a + b$ of the numbers $ab, ab - 1, ab - 2, \ldots, ab - (a + b) + 1$ to be good. You also expect exactly half of the $(a - 1)(b - 1)$ numbers $0, 1, 2, \ldots, ab - (a + b)$ to be good. You would therefore like to show that $N = (a + b) + \frac{1}{2}(a - 1)(b - 1) = \frac{1}{2}(a + 1)(b + 1)$.

Exercise 39 (i) Suppose $c < ab = 20$ is good in two different ways; $c = 5x + 4y = 5u + 4v$. Then $0 \leqslant x, u \leqslant 3$ and $0 \leqslant y, v \leqslant 4$, and $5(x - u) = 4(v - y)$. So 5 would have to divide $v - y$, which is impossible.

(ii) The number $c = 20$ is certainly good in at least two ways: $20 = 5 \times 4 + 4 \times 0 = 5 \times 0 + 4 \times 5$. If 20 were good in some other way, say $20 = 5x + 4y$, then $1 \leqslant x \leqslant 3$ and $1 \leqslant y \leqslant 4$ and $5(4 - x) = 4y$. So 5 would have to divide y, which is impossible.

(v) The lattice points to be counted make up exactly half the lattice points in the rectangle with corners $(0, 0)$, $(4, 0)$, $(4, 5)$, $(0, 5)$ (rotate the triangle of marked points in Fig. 15.4(ii) through $180°$ about the centre of the rectangle to get the other half). Therefore $N = \frac{1}{2}(4 + 1)(5 + 1)$.

Exercise 40 (i) Suppose $c < ab = 30$ is good in two different ways: $c = 5x + 6y = 5u + 6v$. Then $0 \le x$, $u \le 5$ and $0 \le y$, $v \le 4$, and $5(x - u) = 6(v - y)$. So 5 would have to divide $v - y$, which is impossible.

(ii) The number $c = 30$ is certainly good in two different ways: $30 = 5 \times 6 + 6 \times 0 = 5 \times 0 + 6 \times 5$. If 30 were good in some other way, say $30 = 5x + 6y$, then $1 \le x \le 5$ and $1 \le y \le 4$ and $5(6 - x) = 6y$. So 5 would have to divide y, which is impossible.

(iv) The lattice points to be counted make up exactly half the lattice points in the rectangle with corners $(0, 0)$, $(6, 0)$, $(6, 5)$, and $(0, 5)$. Therefore $N = \frac{1}{2}(6 + 1)(5 + 1)$.

Exercise 41 (i) Suppose $c < ab$ is good in two different ways: $c = ax + by = au + bv$. Then $0 \le x$, $u \le b - 1$ and $0 \le y$, $v \le a - 1$, and $a(x - u) = b(v - y)$. So a would have to divide $v - y$ which is impossible.

(ii) The number $c = ab$ is certainly good in two different ways: $ab = a \times b + b \times 0 = a \times 0 + b \times a$. If ab were good in some other way, say $ab = ax + by$, then $1 \le x \le b - 1$ and $1 \le y \le a - 1$, and $a(b - x) = by$. So a would have to divide y which is impossible.

(iii) The lattice points to be counted make up exactly half the lattice points in the rectangle with corners $(0, 0)$, $(b, 0)$, (b, a) and $(0, a)$. Therefore $N = \frac{1}{2}(a + 1)(b + 1)$.

Exercise 42 The hint to Part (i) shows that $\frac{1}{2}(a + 1)(b + 1) = N = $ (number of good numbers c satisfying $0 \le c \le ab - (a + b)$) + (number of good numbers c satisfying $ab - (a + b) < c \le ab) \le \frac{1}{2}(ab - (a + b) + 1) + (ab - (ab - (a + b)) = \frac{1}{2}(a + 1)(b + 1)$, so we must have equality throughout.

(i) In particular the number of good numbers c satisfying $ab - (a + b) < c \le ab$ must in fact be equal to $ab - (ab - (a + b)) = a + b$, so ab, $ab - 1, \ldots, ab - (a + b) + 1$ are all good.

(ii) And the number of good numbers c satisfying $0 \le c \le ab - (a + b)$ must be equal to $\frac{1}{2}(ab - (a + b) + 1) = \frac{1}{2}(a - 1)(b - 1)$. Exercise 36 shows that if c is good, then $ab - (a + b) - c$ is bad; and Exercise 37 now shows that if c is bad, then $ab - (a + b) - c$ is good.

16 The coin problem

The Postage Stamp Problem is rarely mentioned explicity in text-books. But it is closely related to a fundamental problem which appears in every textbook on number theory or abstract algebra (and which holds the key to measuring problems like the one we stated at the very beginning of Chapter 12). It would be a pity to end the investigation without taking a look at this important related problem. So that is what we shall do in this final short chapter.

In a country with 5c and 8c stamps only, all postal rates would have to be good numbers, so one could never charge 22c (say) for a letter. But if the same country had only 5c and 8c *coins*, shopkeepers would not be quite so restricted as the postal service: if an item cost 22c (a bad number) a shopkeeper could take six 5c coins and give one 8c coin as change.

Exercise 43* You are a shopkeeper in a country with only 5c and 8c coins. A price is *chargeable* if a customer with enough coins would be able to pay exactly that amount *provided you were willing to give change.* 0c, 5c, and 8c are obviously chargeable. But so also is 3c (take 8c, give 5c change).

(i) Complete the table in Fig. 16.1.

Amount	0c	1c	2c	3c	4c	5c	6c	7c	8c	9c	10c
Chargeable?											
Not chargeable?											

Fig. 16.1

(ii) Which of the following are chargeable: 749c, 62717c, 7312465978c?

Exercise 44* You move to a new country which has only 5c and 9c coins and set up shop again. The first thing you have to do is to work out from scratch which prices are chargeable *in your new country,* and which are not chargeable. Draw up a table like the one in Exercise 43 and extend it until you think you can say exactly which prices are chargeable and which are not.

Exercise 45* Your next move is to a country that has only 5c and 10c coins. Which prices are chargeable in such a country and which are not?

Exercise 46* Some years later you are on the move again—this time to a country which has only 5c and 11c coins.
    (i) Guess which prices are chargeable in such a country.
    (ii) Draw up a table like the one in Exercise 43 and extend it until you think you can say exactly which prices are chargeable and which are not.

In Exercises 43–46 the value of the smallest coin is always 5c. So as soon as you get *five ticks in a row* you can be absolutely certain that all larger numbers are chargeable (just as in the Postage Stamp Problem).

In Exercise 45 it is clear that the only chargeable prices are multiples of 5c. But the other exercises suggest rather strongly that if a country has a cent and b cent coins only, then *every price is chargeable* provided that the two numbers a and b have no common factors. Translating this guess into algebra, it looks as if you can get any number you like (say N), by handing over a suitable number of a cent coins and a suitable number of b cent coins (say $sa + tb$), and receiving change (say $ua + vb$). But if N cents can be paid by handing over $sa + tb$ and receiving $ua + vb$ as change (where $s, t, u, v \geqslant 0$), then

$$N = (sa + tb) - (ua + vb)$$

so N can be written in the form $N = xa + yb$, where $x = s - u$ and $y = t - v$.

The crucial difference between the Coin Problem and the Postage Stamp Problem is that though s, t, u, v are all $\geqslant 0$, one of the numbers x, y may be <0—so a number may be chargeable without being good. But you should feel that the two problems are sufficiently similar for there to be a reasonable chance of being able to use what you know about the Postage Stamp Problem to prove what you suspect to be true in the Coin Problem.

Exercise 47* Let a and b be positive whole numbers with no common factors. Show that in a country which has only a cent and b cent coins, every whole number N is chargeable.

What is the most important chargeable number? In Exercise 47 you showed that every number N can be expressed in the form $N = xa + yb$, provided that a and b have no common factors. But what matters most (in a sense) is that the number '1' can be expressed in this way. For as soon as you know that $1 = xa + yb$, you can immediately write any other number N in terms of a and b: $N = (Nx)a + (Ny)b$. Unfortunately, though you proved in Exercise 47 that it is always possible to write the number '1' in the form $1 = xa + yb$ by choosing suitable values of x and y, *you have as yet no satisfactory way of finding suitable values of x and y.*

Exercise 48 (i) (*a*) Find whole numbers x and y with $12x + 7y = 1$.

(b) Use Part (*a*) to obtain 1 pint using a 12 pint jug and a 7 pint jug. (The problem is concerned not with money, but with measuring water using two containers of given capacity: 12 pint and 7 pint. However the idea is exactly the same as the Coin Problem with 12c and 7c coins.)

(ii) (a) Find whole numbers x and y with $89x + 55y = 1$.

(b) Use Part (*a*) to obtain 1 pint using an 89 pint jug and a 55 pint jug.

Exercise 48 (ii) makes it painfully clear that, although it may always be *theoretically possible* to find suitable values of x and y, it does not seem to be all that easy in practice. Intelligent trial and error may be the only method available to the beginner, but we would obviously like a simpler and more reliable way of finding suitable values of x and y. We end by describing one such method.

When $a = 12$ and $b = 7$ it does not take long to notice that $3a = 36$ and $5b = 35$, so $3a - 5b = 1$. The same idea can be used when $a = 89$ and $b = 55$, but you have to work out an awful lot of multiples of 89 and test each one in turn. An alternative method is shown in Fig. 16.2 for the example $a = 12$, $b = 7$. It may look a bit long-winded at first sight, but it has two considerable advantages. First it is based on repeated *subtraction*, and subtraction is much quicker and simpler than multiplication. Second it is based on repeating one simple step over and over again, so it is very easy to carry out (or to program a calculator to do the work for you)—something which should become apparent when you use the same method for other pairs a, b in Exercises 49–51.

Exercise 49 Suppose $a = 19$ and $b = 17$. Use the same procedure to find whole numbers x, y satisfying $19x + 17y = 1$.

$$a = 12, b = 7$$

Step 1 The two smallest numbers obtained so far are **12** and **7**. Subtracting the smaller one from the larger one as often as you can, you get $12 - 1 \times 7 = 5$.

Step 2 The two smallest numbers obtained so far are **7** and **5**. Subtracting the smaller one from the larger one as often as you can, you get $7 - 1 \times 5 = 2$.

Step 3 The two smallest numbers obtained so far are **5** and **2**. Subtracting the smaller one from the larger one as often as you can, you get $5 - 2 \times 2 = 1$

Step 4 You have now got **1** by combining **5** and **2**. Substitute for **2** from the equation in Step 2 into the equation in Step 3 to get $5 - 2 \times (7 - 1 \times 5) = 1$. That is, $3 \times 5 - 2 \times 7 = 1$.

Step 5 You have now got **1** as a combination of **7** and **5**. Substitute for **5** from the equation in Step 1 into the equation in Step 4 to get $3 \times (12 - 1 \times 7) - 2 \times 7 = 1$. That is $3 \times 12 - 5 \times 7 = 1$. You have now got **1** as a combination of **12** and **7** as required.

Fig. 16.2

Exercise 50 Suppose $a = 34$ and $b = 21$. Use the same procedure to find whole numbers x, y satisfying $34x + 21y = 1$.

Exercise 51 Suppose $a = 89$ and $b = 55$. Use the same procedure to find whole numbers x, y satisfying $89x + 55y = 1$.

The procedure you have just been using is called *Euclid's Algorithm* because it appears in Euclid's book *The Elements*. The procedure was invented more than two thousand years ago, but it is still an important tool in mathematics and you will probably meet it again.

Hints to Exercises

47. If N is any whole number >0, then you know (for example) that ab and $ab + N$ are both good numbers. So $ab + N = sa + tb$ and $ab = ua + vb$ for some whole numbers s, t, u, $v \geqslant 0$. Hence $N = \ldots$.

Solutions

Exercise 43 Each of the given amounts is chargeable.

Exercise 44 Each of the given amounts is chargeable.

Exercise 45 All multiples of 5c are chargeable: all other amounts are not.

Exercise 46 (i) *Guess*: All prices are chargeable.

(ii) As soon as you find *five* consecutive chargeable numbers you can immediately conclude that all larger numbers are chargeable (exactly as in the Postage Stamp Problem, Exercises 11-13). So to check your guess you only need to show that the first five numbers 0, 1, 2, 3, 4 are all chargeable.

Exercise 47 The hint shows that $N = (s - u)a + (t - v)b$ can be obtained as a combination of as and bs, so N is chargeable.

Exercise 48 There are infinitely many solutions to each of these problems.

(i) (*a*) The most obvious solution here is $12 \times 3 + 7 \times (-5) = 1$ (though once we know that the line $12x + 7y = 1$ passes through the lattice point $(3, -5)$, we know that it must also pass through $(3 + 7m, -5 - 12m)$, where *m* may be any whole number). (*b*) Fill the 12 pint jug 3 times, using its contents to fill up the 7 pint jug 5 times (each time the 7 pint jug is completely full its contents are simply thrown away).

(ii) (*a*) None of the solutions is obvious here since the numbers are rather large. The smallest solution is $89 \times (-21) + 55 \times 34 = 1$. (*b*) Fill the 55 pint jug 34 times, using its contents to fill up the 89 pint jug 21 times (each time the 89 pint jug is completely full its contents are simply thrown away).

Exercise 49 $19 - 1 \times 17 = 2$, $17 - 8 \times 2 = 1$. Hence $17 - 8(19 - 1 \times 17) = 1$, so $19 \times (-8) + 17 \times 9 = 1$.

Exercise 50 $34 - 1 \times 21 = 13$, $21 - 1 \times 13 = 8$, $13 - 1 \times 8 = 5$, $8 - 1 \times 5 = 3$, $5 - 1 \times 3 = 2$, $3 - 1 \times 2 = 1$. Hence

$$\begin{aligned}
1 &= 3 - 1 \times 2 = 3 - 1(5 - 1 \times 3) = (-1) \times 5 + 2 \times 3 = (-1) \times 5 + 2(8 - 1 \times 5) \\
&= (-3) \times 5 + 2 \times 8 = (-3)(13 - 1 \times 8) + 2 \times 8 = 5 \times 8 - 3 \times 13 \\
&= 5(21 - 1 \times 13) - 3 \times 13 = 5 \times 21 - 8 \times 13 = 5 \times 21 - 8(34 - 1 \times 21) \\
&= 13 \times 21 + (-8) \times 34.
\end{aligned}$$

Exercise 51 $89 - 1 \times 55 = 34$, $55 - 1 \times 34 = 21$, $34 - 1 \times 21 = 13, \ldots$. You are now simply repeating the steps you took in Exercise 50, so you can use what you found there, namely

$$\begin{aligned}
1 &= 13 \times 21 + (-8) \times 34 = 13(55 - 1 \times 34) + (-8) \times 34 = 13 \times 55 + (-21) \times 34 \\
&= 13 \times 55 + (-21)(89 - 1 \times 55) = (-21) \times 89 + 34 \times 55.
\end{aligned}$$

Postscript

In our investigation of the Postage Stamp Problem we have focussed exclusively on the question: Which amounts can be made using just *two* kinds of stamps—*a* cent and *b* cent? But there is no obvious reason why one should restrict attention to this two stamp problem. Could one not ask exactly the same question when *three, four*, or *n* different kinds of stamps are available? It was probably a good idea to consider just two kinds of stamps to begin with. But now that you have found several different ways of looking at the Postage Stamp Problem for two kinds of stamps, you should be in a good position to start investigating harder versions of the same problem for yourself.

One particularly interesting variation of the two stamp problem is the game 'Sylver Coinage', which is discussed in Chapter 18 of the book *Winning Ways* by Berlekamp, Conway, and Guy (published by Academic Press in 1982). But this Postscript leads in a different direction in the hope that you might begin exploring the *three* stamp problem on your own.

Suppose you have an inexhaustible supply of *three* kinds of stamps: say *a* cent, *b* cent, and *c* cent. Which amounts can be obtained using just these stamps? Is there a last bad number? If so, how does it depend on the values of *a*, *b*, and *c*?

It will be convenient to assume that *a* is the smallest of the three numbers *a*, *b*, *c* and that *c* is the largest. In this version of the Postage Stamp Problem a number *d* will be *good* if it can be written in the form $d = ax + by + cz$ (where *x, y, z* are whole numbers ≥ 0). If the three numbers *a*, *b*, *c* have a common factor *f*, then every good number $ax + by + cz$ will automatically be a multiple of *f*. It therefore seems reasonable to assume that 1 *is the only number which divides all three of the numbers a, b, c*. Here are some exercises to get you started.

Exercise 52 Suppose $a = 2 \leq b \leq c$ (and that 1 is the only number which divides all three of the numbers *a*, *b*, *c*).
   (i) Show that either *b* or *c* is odd.
   (ii) If *b* is odd, show that $b - 2$ is the last bad number.

(iii) If b is even (so c is odd), show that $c - 2$ is the last bad number.

Exercise 53 Suppose $a = 3 \leqslant b \leqslant c$.

(i) Suppose that b is a multiple of 3 (so c is not). Is there a last bad number? If so, what is it?

(ii) Suppose that c is a multiple of 3 (so b is not). Is there a last bad number? If so, what is it?

(iii) Suppose that neither b nor c is a multiple of 3. Find a formula for the last bad number.

Exercise 54 Suppose $a = 4$.

(i) Suppose that b is even (so c is odd). Is there a last bad number? If so, what is it?

(ii) Suppose that c is even (so b is odd). Is there a last bad number? If so, what is it?

(iii) Suppose that b and c are both odd. What is the last bad number?

Exercise 55 Suppose that a, b, c are any three numbers with $a \leqslant b \leqslant c$, and suppose that 1 is the only number which divides all three of a, b, c. Find a whole number G such that every whole number $\geqslant G$ is good.

Your solution to Exercise 55 shows that there is always a last bad number B, and that $B < G$. But the actual value of B is usually much smaller than the number G which you found in Exercise 55.

Exercise 56 (i) Find B when $a = 5$, $b = 6$, $c = 7$. Compare this value of B with the value of G which you found in Exercise 55, (and with the value $ab - (a + b)$ which would be the last bad number if you used only a cent and b cent stamps).

(ii) Do the same when $a = 5$, $b = 6$, $c = 8$.

(iii) Do the same when $a = 5$, $b = 6$, $c = 9$.

Exercise 57 (i) Find B when $a = 6$, $b = 10$, $c = 15$. Compare this value of B with the value of G which you found in Exercise 55.

(ii) Do the same when $a = 7$, $b = 8$, $c = 9$.

(iii) Do the same when $a = 7$, $b = 10$, $c = 13$.

Exercises 52–7 suggest that the formula for the last bad number B is unlikely to be as simple-looking as the formula $ab - (a + b)$ for the last bad number in the two stamp problem. You could start tracking down a formula for the exact value of B by looking back at what you found in Exercises 52–4 and trying first to find such a formula when $a = 2$, then when $a = 3$, then when $a = 4$, and so on. There is plenty of scope here for experiment and calculation. But if you fail to find a formula for the exact value of B, you could go back to Exercise 55 and look for ways of improving your value of G to bring it nearer to the actual value of B. (After you have developed your own approach to the problem you might like to look up the article 'Skolem's solution to a problem of Frobenius' by C. Smoryński on pages 123–32 of *The Mathematical Intelligencer*, Vol. 3, No. 3, 1981—published by Springer Verlag.)

Hints to Exercises

53. (i) Since $a = 3$ and b is a multiple of 3, you can simply ignore b. So the question is a disguised version of the Postage Stamp Problem with two types of stamps: a cent and c cent.
   (ii) Read the hint for Part (i).
   (iii) If $c - b$ is a multiple of 3, then you can ignore c and concentrate on a and b only. So there are really two separate cases: (a) $c - b$ is a multiple of 3, and (b) $c - b$ is not a multiple of 3.
54. (i) Either (a) $b = 4b'$ is a multiple of 4, or (b) $b = 4b' + 2$ is not a multiple of 4. Consider these two cases separately.
   (ii) Read the hint for Part (i).
   (iii) If $c - b$ is a multiple of 4, then you can ignore c. If $c - b$ is even, but not a multiple of 4, then either $c > 3b - 4$ or $c \leqslant 3b - 4$. Consider these two cases separately.
55. You know all about the Postage Stamp Problem with two types of stamps. Apply this knowledge first to the pair a, b; then to the pair . . . . (But beware! The two numbers a, b may have a common factor.)
56. (i) Draw up a table like the one in Exercise 5 and check each number until you find 5 ($=a$) good numbers in a row.

Solutions

Exercise 52 (i) $a = 2$ and a, b, c cannot all be multiples of 2, so at least one of b, c must be odd.
   (ii) Since $a < b < c$, the only good numbers $< b$ are multiples of a. If $a = 2$ and b is odd, then $b - 2$ is odd and hence bad, $b - 1$ is even and hence good, and b is obviously good. So $b - 2$ is the last bad number.

(iii) If b is even, then c is odd and each good number can be obtained using a and c only. Your work on the Postage Stamp Problem for two kinds of stamps then shows that $ac - (a + c) = c - 2$ is the last bad number.

Exercise 53 (i) If b is a multiple of 3, then c is not a multiple of 3 and each good number can be obtained using a and c only. Hence $ac - (a + c) = 2c - 3$ is the last bad number.

(ii) If c is a multiple of 3, then b is not a multiple of 3 and each good number can be obtained using a and b only. Hence $ab - (a + b) = 2b - 3$ is the last bad number.

(iii) (a) If $c - b$ is a multiple of 3, then each good number can be obtained using $a(=3)$ and b only. Hence $ab - (a + b) = 2b - 3$ is the last bad number. (b) If $c - b$ is not a multiple of 3 there are two possibilities: either $c > ab - (a + b) = 2b - 3$, or $c \leqslant 2b - 3$. In the first case c itself can be obtained as a combination of as and bs and so can be ignored: the last bad number is therefore $2b - 3$. In the second case c cannot be obtained as a combination of as and bs (if $c = ax + by$, then $y \leqslant 1$; but $y \neq 0$ since c is not a multiple of 3, and $y \neq 1$ since $c - b$ is not a multiple of 3), so neither can $c - 3$. Hence $c - 3$ is bad. But one of $c - 2$, $c - 1$ is a multiple of 3 and the other is equal to 'b plus a multiple of 3', so $c - 2$ and $c - 1$ are both good. Since c is obviously good, $c - 3$ is the last bad number.

Exercise 54 (i) (a) If $b = 4b'$ is a multiple of $a = 4$, then each good number can be obtained using a and c only, so the last bad number is $ac - (a + c) = 3c - 4$. (b) If $b = 4b' + 2$ is not a multiple of $a = 4$, then every even number $\geqslant b - 2$ is good. Similarly every odd number $\geqslant b + c - 2$ is good. So every number $\geqslant b + c - 3$ is good. It is not hard to show that $b + c - 4$ is bad. Hence $b + c - 4$ is the last bad number.

(ii) If c is even, then b is odd: say $b = 2b' + 1$. Hence $2b = 4b' + 2$, so every even number $\geqslant 2b - 2$ is good. (a) If $c = 4c'$ is a multiple of $a = 4$, then each good number can be obtained using a and b only, so $ab - (a + b) = 3b - 4$ is the last bad number. (b) If $c = 4c' + 2$ is not a multiple of $a = 4$, then every even number $\geqslant c - 2$ is good. $3b$ is obviously good, $3b - 1$ and $3b - 3$ are even and hence good, and $3b - 2 = b + (2b - 2)$ is also good. If $c \geqslant 2b$, then $3b - 4$ is bad, so $3b - 4$ is the last bad number. If $c < 2b$, then $c \leqslant 2b - 4$, every odd number $\geqslant b + c - 2$ is good, and $b + c - 4$ is bad; hence $b + c - 4$ is the last bad number.

(iii) Every even number $\geqslant 2b - 2$ is good, and $2b - 4$ is bad. Also every odd number $\geqslant 3b - 2$ is good. (a) If $c - b$ is a multiple of 4, then every odd number $\geqslant c - 2$ is good. So if $c \geqslant 3b$, then $3b - 4$ is the last bad number. But if $c < 3b$, then $c - 4$ is the last bad number.

Exercise 55 There are lots of possible answers. Here is one. If a and b have no common factors, then you could choose $G = (a - 1)(b - 1)$. Similarly if b and c, or c and a have no common factors. In general, let d be the highest common factor of a and b. Then $a = a'd$, $b = b'd$ and a', b' have no common factors. Hence every number $\geqslant (a' - 1)(b' - 1)$ can be written in the form $a'x + b'y$ ($x, y \geqslant 0$), so every multiple of d which is $\geqslant (a' - 1)(b' - 1)d$ can be written in the form $ax + by$ ($x, y \geqslant 0$). Let ed be the first such multiple of d

which has no factors in common with c ($ed < (a' - 1)(b' - 1)d + cd$, because d and c have no common factors). Then every number $\geqslant (ed - 1)(c - 1)$ can be written in the form $ew + cz$ ($w, z \geqslant 0$), and hence in the form $ax + by + cz$ ($x, y, z \geqslant 0$). So we may choose $G = (ed - 1)(c - 1)$.

Exercise 56 (i) 9 is bad and 10, 11, 12, 13, 14 are good; so $B = 9$ is the last bad number. (If you use only $a = 5$ and $b = 6$, then $ab - (a + b) = 19$ is the last bad number, and $G = 20$. If you use my solution to Exercise 55, you get $d = 1$, $(a' - 1)(b' - 1) = 20 = ed$, $G = 19 \times 6 = 114$.)

(ii) 9 is bad and 10, 11, 12, 13, 14 are good, so $B = 9$ is the last bad number. (If you use only $a = 5$ and $b = 6$, then $G = 20$ as in (i). If you use my solution to Exercise 55, then $d = 1$, $(a' - 1)(b' - 1) = 20$, $ed = 21$, and $G = 20 \times 7 = 140$.)

(iii) 13 is bad and 14, 15, 16, 17, 18 are good, so $B = 13$ is the last bad number. (If you use only a and b, then $G = 20$ as before. If you use my solution to Exercise 55, then $d = 1$, $(a' - 1)(b' - 1) = 20 = ed$, and $G = 19 \times 8 = 152$.)

Exercise 57 (i) $B = 29$, $G = 15 \times 14$ (since $d = 2$, $a' = 3$, $b' = 5$, $(a' - 1)(b' - 1) = 8$, $ed = 16$).

(ii) $B = 20$, $G = 41 \times 8$ (since $d = 1$, $a' = 7$, $b' = 8$, $(a' - 1)(b' - 1) = 42$, $ed = 42$).

(iii) $B = 32$, $G = 53 \times 12$ (since $d = 1$, $a' = 7$, $b' = 10$, $(a' - 1)(b' - 1) = 54$, $ed = 54$).

17 Further problems for extended investigation

*The best way of overcoming a difficult Probleme is
to solve it in some particular easy cases. This
gives much light into the general solution. By
this way Sir Isaac Newton says he overcame the
most difficult things.*

(David Gregory, *Memorandum* 1705(?))

17.1 Introduction

What is Mathematics? This book has addressed this old question from
an unusual point of view—focussing not on the most impressive
achievements of the best mathematicians, but on elementary problems for you to tackle for yourself. The mathematics you have used
and discovered along the way may have been less significant; but it
has, I hope, given you a clearer idea as to where mathematics 'comes
from', and how it is 'discovered'.

One of the two recurring themes of the book has been that
insight—whether it be insight into a specific mathematical problem, or
insight into general philosophical questions such as 'How is mathematics discovered?'—has its roots in *activity* (calculation, reasoning,
picturing, etc.). The other recurring theme has been that activity
needs a guide, and the best guide is *reflection*. If you calculate without
reflecting on what you have been doing, why you have been doing it,
and what you have found, then your activity is likely to lose its sense
of purpose and direction. If on the other hand you try to speculate
without getting involved in messy calculations, you are likely to lose
touch with reality.

For most readers the experience of grappling with extended
investigations will have been new. I have therefore avoided preaching
too many methodological sermons on 'how one *does* mathematics'.
Nevertheless a few general principles should have emerged fairly
clearly, and this may be a good place to summarize some of the more
obvious strategies we have used.

In each of the six examples in the text we began by *experimenting,*
or *calculating,* in particularly easy cases. This usually helped in two
ways—one methodological, the other psychological. First, these

calculations provided us with 'experimental data' which often turned out to be useful later on. But no less important was the fact that these simple-minded initial calculations helped you to understand the problem better. The methods used in these initial calculations are bound to be a bit crude: you should not let that worry you—they can always be improved later.

I have also tried to stress that it is important, at some stage, to step back and *reflect* on the initial evidence: if possible you should try to make an intelligent *guess* as to what seems to be going on; you should then *test your guess* in some way. However even if no pattern is visible, it is important not to go on calculating unthinkingly, but to use your experience to *decide what to do next*. Above all remember that you are unlikely to discover what is going on unless you explore the problem in some kind of *systematic* way.

In a book such as this readers will naturally expect some guidance about how to tackle the six Further problems in this chapter—though since different people will approach the same problem in very different ways, they would hardly expect 'hints' to suit every approach. When you are exploring a problem on your own you will find that many of the lines of thought which you try to develop run into difficulty sooner or later. Some of these are genuine dead-ends. But others would almost certainly bear fruit if only you were to pursue them. You will have to use your own judgement as to which is which.

In order not to rob you of your chance to tackle each problem in your own way I have deliberately separated the *statement* of the Further problems (in Section 17.2) from my own *outline*, with hints, of one possible approach to each problem (in Section 17.3). If you are just browsing through the book, or are about to begin an extended investigation for yourself, then you should **read Section 17.2 only.** Only when you have tried to explore one of the problems in Section 17.2 on your own and have got well and truly stuck should you consult my outline in Section 17.3. The outlines are much less detailed than the text of Chapters 1–16, and much more of the work is left for you to plan, carry out, and check for yourself. There are also fewer clues as to what you can expect to find. However each outline consists of a sequence of steps, and later steps inevitably give away some of the things you are supposed to have discovered for yourself at an earlier stage. So when you do follow my outline, try to complete each step before going on to the next. I have provided hints for some of these steps: these are printed at the end of the outline, and before the outline to the next problem.

I have tried to choose problems which are elementary yet rich in ideas and surprises. There are many such problems in the popular literature, but I have tried to avoid simply reproducing well-known and well-worn examples in the hope that seasoned teachers and lecturers may experience afresh the same kind of challenge as their students. There are no official solutions. The aim is that you should explore, discover, and explain as much as you can on the basis of the initial problem, not that you should feel constrained to try to rediscover some official solution.

Zarathustra's last, most vital lesson: 'now do without me'.
(George Steiner, *Humane Literacy*)

17.2 Further problems

1. A well-known problem asks how many different ways there are of laying n paving stones, each measuring 2×1, on a section of path which measures $2 \times n$. How many different ways are there of paving a $3 \times n$ section of path with 2×1 paving stones?

2. Which terms of the Fibonacci sequence $0, 1, 1, 2, 3, 5, 8, \ldots$ are divisible by n?

3. Which triangles and quadrilaterals can one construct using all seven Tangram pieces? Which convex polygons can you construct?

4. The traditional Towers of Hanoi puzzle has three pegs, and a tower of rings with decreasing size on one of the pegs. The challenge is to find the least number of moves required to transfer this tower onto another peg, moving one ring at a time and never placing a larger ring on top of a smaller one. What is the least number of moves if there are more than three pegs?

5. When travelling from A to B in 'Taxicab geometry' you are only allowed to move parallel to the axes; so you have to go along-and-up (or up-and-along, or . . .) rather than travel-as-the-crow-flies. The shortest distance between $A = (x_1, y_1)$ and $B = (x_2, y_2)$ is therefore

$$\text{distance}\,((x_1, y_1), (x_2, y_2)) = |x_1 - x_2| + |y_1 - y_2|.$$

(a) What loci do the familiar 'focus-directrix' definitions of conics give in Taxicab geometry? What loci do the familiar 'focus-focus' definitions of conics give in Taxicab geometry? Are the two families related? (If so, how? If not, why not?)

(b) Are these loci related in some way to the plane cross-sections of a (Taxicab) circular cone?

6. Is there a general pattern governing flips in any base?

17.3 Outlines and Hints

READ SECTION 17.1 BEFORE LOOKING AT
THE CONTENTS OF THIS SECTION

> 1. A well-known problem asks how many different ways there are of laying n paving stones, each measuring 2×1, on a section of path which measures $2 \times n$. How many different ways are there of paving a **$3 \times n$** section of path with 2×1 paving stones?

Outline

(1.1) First make sure you can find all possible ways of paving a $2 \times n$ section of path with 2×1 paving stones. (Experiment with small values of n—see Fig. 17.1. Then try to find a good way of counting the total number of different ways for a general value of n. Finally try to get a formula for the number of different ways in terms of n alone.)

n	0	1	2	3	4
Number of different ways of paving $2 \times n$ path with 2×1 stones					

Fig. 17.1

(1.2) Now start exploring the problem of paving a **$3 \times n$** section of path with 2×1 paving stones. As in (1.1) you will need a short way of referring to 'the number of different ways of paving a $3 \times n$ section of path with 2×1 paving stones'—say P_n.

One good way of getting a feeling for a problem is to begin experimenting with easy cases. So start by working out the exact value of P_n for small values of n—see Fig. 17.2.

(1.3) When you try to pave a $3 \times n$ section of path for odd values of n you have to leave at least one 1×1 hole. This is a bit like trying to pave the path with one paving stone measuring 1×1, and $(3n - 1)/2$ paving stones measuring 2×1. This may be an interesting problem for all we know, but it looks rather different from the case where n is even and one has to fill the $3 \times n$ section of path with 2×1 paving

Length of path = n	0	1	2	3	4	5	6
'Number of different ways of paving a $3 \times n$ path with 2×1 stones' = P_n							

Fig. 17.2

stones only. So it seems reasonable to stick to the case where n is even—at least to begin with. You can then concentrate on the values of P_n which you found in (1.2) for even n (Fig. 17.3).

n	0	2	4	6
P_n				

Fig. 17.3

(1.4) If you managed to solve (1.1) satisfactorily, with or without looking at the hint, then you presumably agree that $P_0 = 1$. It is not hard to see that $P_2 = 3$: all three stones 'along' the path (one way), or one stone 'along' the path and two stones 'across' it (two ways). Finding P_4 is already more complicated. And if you did not devise some clever way of counting, then you will have had considerable difficulty working out P_6. In that case you cannot be sure that the value you found in (1.2) is correct. So though it seems like hard work it is a good idea to try to work out P_6, P_8 and P_{10} exactly—if only to force yourself to find some better way of counting. Do this now. (An improved method of counting is obviously needed if you are ever going to find a clever way of working out P_{500}, or P_{1000}, or P_{2m} for general m.)

(1.5) Examine the method you used in (1.4) to work out P_6, P_8, P_{10}, and try to use the same idea to write down a recurrence relation which will allow you to calculate P_n (n even) as soon as you know $P_{n-2}, P_{n-4}, \ldots, P_2, P_0$.

(1.6) Your first recurrence relation will probably be a bit complicated. But then so was the first recurrence relation you found for the number of 9-flips with n digits (Exercise 39, Chapter 9)! Try to simplify the recurrence relation you found in (1.5) in the same sort of way as you simplified the recurrence relation for the number of 9-flips with n digits (Exercise 45, Chapter 9).

(1.7) Then use the same method as in Chapter 9 to find a formula for P_n (n even) in terms of n alone.

(1.8) It is always tempting to abandon a problem as soon as you feel you have managed to solve it successfully. But there are still several interesting loose ends. For example:
  (i) The $3 \times n$ problem originally arose as a natural generalization of another problem (paving a $2 \times n$ section of path). How are the two problems related?
  (ii) You postponed investigation of the problem of paving a $3 \times n$ section of path when n is odd.

(1.9) So go back to the $2 \times n$ problem *for even values of n*. Let p_n be the number of ways of paving a $2 \times n$ path with 2×1 stones. Use your solution to (1.1) to fill in the table in Fig. 17.4.

n	0	2	4	6	8	10	12
p_n							

Fig. 17.4

Try to find a recurrence relation for p_n (n even) which 'fits in' in some way with either the first recurrence relation you found in (1.5), or the simpler recurrence relation you found in (1.6). What does this suggest about the number of different ways of paving a $4 \times n$ section of path with 2×1 paving stones (when n is even)?

(1.10) Check your guess about the recurrence relation for the number of ways of paving a $4 \times n$ section of path with 2×1 paving stones (when n is even). If it seems to be correct, try to prove it. If not, try to revise your guess to find a recurrence relation which is correct.

(1.11) You could then go back to consider the problem of paving a $3 \times n$ path when n is odd (using $(3n-2)/2$ paving stones, each measuring 2×1).

Hints

(1.1) You should by now realize that you will need a short way of referring to 'the number of different ways of paving a $2 \times n$ section of path with 2×1

paving stones', say p_n (**p** for **p**aving). This is a fairly easy problem—experimenting with small values of n will probably suggest a fairly convincing 'first guess' as to what p_n is in general. (However you will have to think hard about what at first sight looks like the easiest case. *How many ways are there of paving a 2×0 section of path with 2×1 paving stones?* There are those who would say 'It can't be done, so the number of ways in which it can be done must be *zero*'. But others might equally well claim 'there is precisely *one* way of paving a 2×0 section of path: namely, sit down and do absolutely nothing'. Which of these two points of view do you think a mathematician would accept? Which of them fits in best with the values you found for $p_1, p_2, p_3, p_4 \ldots$?)

You must then find some clever way of working out p_n for general values of n, which will show whether your first guess was in fact correct.

(1.4) If you managed to solve (1.1) satisfactorily you will already be looking for ideas that might eventually help you find a simple *recurrence relation* for P_n. If you were reasonably careful in (1.2) then you should be fairly confident that your value for P_4 is correct. So think again about how you could calculate P_4 very simply by making use of the fact that you already know that $P_2 = 3$ (and $P_0 = 1$). There are two very different ways of paving a 3×4 section of path. One of these ways is closely related to P_2; the other is entirely different. See if you can identify the two different ways for yourself and count the number of distinct pavings of each type before reading on.

A method of paving which is closely related to P_2 must presumably involve a 3×2 section of path. This suggests that one should distinguish between (i) those pavings which have a 'fault line' splitting the 3×4 section of path into two separate 3×2 sections, and (ii) those pavings which cannot be split up in this way. Count each type separately. The combined total should give you P_4. Now try to use the same idea to work out P_6, P_8, and P_{10}.

(1.9) Suppose that n is even. You know that $p_{n+2} = p_{n+1} + p_n$, $p_{n+1} = p_n + p_{n-1}$, and that $p_n = p_{n-1} + p_{n-2}$. Can you combine these to get a recurrence relation expressing p_{n+2} in terms of p_n and p_{n-2}? What is the corresponding recurrence relation expressing P_{n+2} in terms of P_n and P_{n-2}? What does this suggest for the $4 \times n$ problem?

(1.10) Remember that your guess about the recurrence relation for the $4 \times n$ problem (with n even) was a rather optimistic first guess. So you should not be surprised to find either that it needs to be modified in some way, or that it has to be radically revised.

To check your first guess you might start by carefully calculating the first few entries in the table in Fig. 17.5. (Make use of what you already know about the number of ways of paving $2 \times n$ and $3 \times n$ paths.)

n	0	1	2	3	4	5	6
Number of ways of paving a $4 \times n$ path with 2×1 stones							

Fig. 17.5

Your first concern may be to test your guessed recurrence relation for the number of ways of paving a $4 \times n$ path when n is *even*. But the number of ways of paving a $4 \times n$ path for the intermediate *odd* values of n may be useful in various ways: for example, you may find them helpful when you try to work out the number of ways of paving a 4×4 path or a 4×6 path. Try to fill in these values before reading on.

The first test of your guessed recurrence relation comes as soon as you have worked out the number of ways of paving a 4×0, a 4×2 and a 4×4 path. The first of these values is presumably the same as for a 2×0 and a 3×0 path. The second of these values is presumably the same as the value you found for a 2×4 path in (1.1). Only the 4×4 path presents any problems. But the total number of ways of paving such a path certainly includes those which have a 'fault line' across the middle of the path splitting it into two separate 4×2 pieces: and the number of these special pavings is probably already big enough to show that your guessed recurrence relation is wrong. (This might appear particularly unfortunate if you have already noticed that a similar recurrence relation is satisfied not only for $2 \times n$ paths and $3 \times n$ paths, but also for the number of ways of paving a $1 \times n$ path with 2×1 stones when n is even. But this observation is irrelevant, because in this case there is always precisely one way of paving the path and the sequence $1, 1, 1, 1, \ldots$ can be made to satisfy all sorts of recurrence relations. The simplest recurrence relation for the $1 \times n$ problem (namely $P_{2n} = 1$) involves just one term of the sequence. The simplest recurrence relations for the $2 \times n$ and $3 \times n$ problems (n even) involve exactly three terms of each sequence. So perhaps one should expect the simplest recurrence relation for the $4 \times n$ problem to be more complicated—perhaps involving five terms of the sequence.)

Where should you go from here? One possibility is to persist with the problem of trying to find a correct recurrence relation for the $4 \times n$ problem when n is even. Another possibility is to decide that the failure of your guessed recurrence relation for the $4 \times n$ problem when n is even may reflect the fact that it was slightly perverse to separate the $2 \times n$ problem for even n from the $2 \times n$ problem for odd n in the first place. And if the $2 \times n$ problem is best considered for all values of n simultaneously, then perhaps the same is true of the $4 \times n$ problem. Once you decide to do this it is not hard to get a recurrence relation for P_n (= the number of ways of paving a $4 \times n$ path) provided one is prepared to introduce two other sequences as 'stepping stones', like the sequence i_n in the Flips problem (Fig. 17.6).

Fig. 17.6

One can then get three recurrence relations involving the three sequences, and can try to eliminate the terms corresponding to the two auxiliary sequences to get a recurrence relation involving the P_ns alone (in fact involving exactly five terms of the P_n sequence).

Since (1.10) started from your first guess at a recurrence relation for the number of pavings of a $4 \times n$ path when n is even, it may be worth combining the recurrence relations for P_n, P_{n-1}, P_{n-2}, P_{n-3}, P_{n-4} to eliminate P_{n-1}, P_{n-3}, P_{n-5}, P_{n-7} and so obtain a recurrence relation involving only P_n, P_{n-2}, P_{n-4}, P_{n-6}, P_{n-8}.

(**1.11**) It may help to get started if you think of your $3 \times n$ section of path as made up of black and white squares like a chessboard. This should tell you where the $1 = 1$ hole can and cannot occur.

Let S_{2n+1} stand for the number of ways in which a $3 \times (2n + 1)$ path can be paved using $3n + 1$ stones (leaving one 1×1 hole). Find a systematic way of calculating S_{2n+1} when $n = 0, 1, 2, 3$. Then try to find a recurrence relation for S_{2n+1}. (This is not too hard provided you are prepared to start by producing a recurrence relation which involves the two 'stepping stone' sequences P_{2n-2} and T_{2n-1}, where P_{2n-2} is the old sequence of (1.2) and T_{2n-1} is the number of ways of paving the path in Fig. 17.7.)

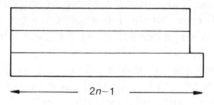

$2n-1$

Fig. 17.7

> 2. Which terms of the Fibonacci sequence 0, 1, 1, 2, 3, 5, 8, . . . are divisible by n?

Outline

(**2.1**) (*a*) Which terms of the Fibonacci sequence 0, 1, 1, 2, 3, 5, 8, . . . are *even*? Experiment, guess, check, . . . , and then prove that your guess is correct.

(*b*) Will the 9000th term be even? What about the 9001st term?

(**2.2**) Your answer to (2.1)(*a*) would have been slightly simpler if you had been willing to think of the term '0' at the beginning of the Fibonacci sequence as the *zero*th term F_0: so $F_0 = 0$, $F_1 = 1$, $F_2 = 1$, $F_3 = 2$, and so on. *Stick to this notation for the rest of this investigation.*

(*a*) Reword your answer to (2.1)(*a*) in this new notation. Which terms of the Fibonacci sequence $F_0 = 0$, $F_1 = 1$, $F_2 = 1$, $F_3 = 2$, . . . are divisible by 2?

(*b*) Which terms are divisible by 3? Experiment, guess, check, . . . , and then prove that your guess is correct.

(*c*) Which terms are divisible by 5?

(**2.3**) (*a*) Which terms are divisible by 6? Explain. (Use (2.2).)

(*b*) Which terms are divisible by 10? Explain.

(*c*) Which terms are divisible by 15? Explain.

(*d*) If $n = p \times q$ is the product of two different primes p and q, you should be able to tell which terms are divisible by n as soon as you know which terms are divisible by p and which terms are divisible by q. Why is this enough?

(*e*) Which terms are divisible by 5? And which terms are divisible by 7? So which terms do you expect to be divisible by 35? Check whether your prediction is correct.

(**2.4**) (*a*) Suppose that, for each whole number n, you want to be able to tell which terms of the Fibonacci sequence are divisible by n. Why is it enough to concentrate on the case where n is a prime power?

(*b*) Which terms are divisible by 2? by 2^2? by 2^3? by 2^4? Can you see any pattern here? Which terms do you expect to be divisible by 2^5? Check whether your prediction is correct.

(*c*) 2 is often a slightly exceptional prime. So have a look at powers of 3. Which terms are divisible by 3? by 3^2? by 3^3? Can you see any pattern here?

(*d*) Which terms are divisible by 5? by 5^2? Which terms are divisible by 7? by 7^2?

(**2.5**) Steps (2.4)(*c*) and (2.4)(*d*) offer some hope that you might be able to manage prime powers if only you could tell which terms were divisible by primes (though you probably cannot as yet explain the pattern you found in (2.4)).

A Fibonacci number F_n is divisible by 2 precisely when n is divisible by 3: we shall say that multiples of 2 occur in the Fibonacci sequence with *period* 3. Complete the table in Fig. 17.8 as far as the prime $p = 37$.

Prime p		2	3	5	7	11	13	17	19	23	29	31	37
Period													
Relationship between p and its period													

Fig. 17.8

One particular prime stands out as being entirely special—though why is not at all clear at this stage. The other primes fall naturally into families. Try to group the primes into what look like their natural families and describe each family. Use what you have found to predict what you expect for $p = 43$ and $p = 47$. Then check whether your prediction was correct.

(2.6) Now test the primes $p = 61$, $p = 89$, $p = 233$. These will probably force you to modify your idea about what natural families the primes fall into. But there still appear to be two distinct types (excluding the one entirely special prime you noticed in (2.5)).

(2.7) At first you will probably only be in a position to decide which primes p belong in which family, and this will not be enough to tell you exactly which terms of the Fibonacci sequence are divisible by p. Continue refinining and sharpening your guess until you have something you believe to be true. Then try to prove that it is true.

(2.8) When you have got as far as you can with your investigation of the period of multiples of p, where p is prime, go back and have a serious look at multiples of prime powers. If p^r is a power of the prime p, try to decide whether there really is a simple connection between the period of multiples of p and the period of multiples of p^r as you began to suspect in (2.4). If the simple connection seems to be generally true, try to prove it. If you suspect either that this apparently simple connection is not true in general, or that it is too hard to prove, modify the obvious first guess from (2.4) until you get something which you think is generally true and which you can prove. Then prove it.

Hints

(2.1) (a) 0, 1, 1, **2**, 3, 5, **8**, 13, 21, **34**, 55, 89, **144**, . . .
Suppose you know that the first two terms of a sequence are even and odd respectively, and that each succeeding term is obtained by adding the previous two terms. How does the sequence continue?

even, odd, ____, ____, ____, ____, ____, ____, . . .

(2.2) (b) Suppose you know that the first two terms of a sequence are a 'multiple of 3' and a '(multiple of 3) + 1' respectively, and that each succeeding term is obtained by adding the previous two terms. Then the next term must be a '(multiple of 3) + 1', the one after that a '(multiple of 3) + 2', and so on. How does the sequence continue?

multiple of 3, (multiple of 3) + 1, (multiple of 3) + 1, (multiple of 3) + 2, . . .

(2.4) (b) It is a curious feature of the Fibonacci sequence that it cannot tell the difference between $2^2 = 4$ and $2^3 = 8$. Any term which is divisible by 4 is automatically divisible by 8. (Can you prove this?) It is not clear at this stage whether this can happen for other powers of 2 as well. But it certainly looks as though the 'period' of multiples of 2^2 is related to the period of multiples of 2 in the same way that the period of multiples of 2^4 is related to the period of multiples of 2^3.

(2.7) You will probably have to be content with less than the whole truth—at least in the first instance. Thus you believe that if a prime is in one family then its period divides one function of p, whereas if it is in the other family then its period divides another function of p. Perhaps this is the first thing to try to prove. You can come back and try to pin down the exact periods later on.

(2.8) Suppose $F_n = kp$ is a multiple of p. Then $F_{2n}, F_{3n}, \ldots, F_{pn}$ are all multiples of p. Look at the terms of the Fibonacci sequence modulo p^2. (*a*) What do you notice about the terms $F_{n-1}, F_{2n-1}, F_{3n-1}, \ldots, F_{pn-1}$? (Try some small values of p and see?) If $F_{n-1} \equiv a \pmod{p^2}$, what do you think you can say about $F_{pn-1} \pmod{p^2}$? (*b*) What do you notice about the terms $F_{n+1}, F_{2n+1}, F_{3n+1}, \ldots, F_{pn+1}$? (Try some small values of p and see!) What do you think you can say about $F_{pn+1} \pmod{p^2}$? (Remember that we are assuming $F_n = kp$.) (*c*) So what do you think you can say about $F_{pn} \pmod{p^2}$? (*d*) Now go back and see how much of this you can *prove*. (*e*) Suppose $F_{n'} = k'p^2$. Can you say anything about $F_{pn'} \pmod{p^3}$?

---

3. Which triangles and quadrilaterals can one construct using all seven Tangram pieces? Which convex polygons can one construct?

(3.1) Draw a 10 cm by 10 cm square on stiff card. Mark it as in Fig. 17.9(i). Then cut out your own seven piece Tangram. In the traditional Tangram puzzle you are given the outline of some shape and must then try to fit the seven pieces together in some way to make the given shape. The idea is extremely simple, but the puzzles

Outline

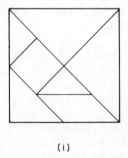

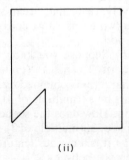

(i)   (ii)

Fig. 17.9

can be surprisingly difficult (Try to make the shape in Fig. 17.9(ii)!)
There are plenty of good books listing interesting shapes which can be
made. But what mathematics is there to *investigate*?

There are so many different ways of arranging the seven pieces that
there is little point trying to describe all possible shapes. But if one
concentrates on particular families of shapes then the possibilities are
far more restricted. The seven pieces are all polygons and the shapes
which can be made by fitting them together are all essentially
polygons. And the simplest polygons are triangles and quadrilaterals.
So it seems reasonable to begin by asking which triangles and
quadrilaterals can be constructed.

(3.2) Find all possible triangles which can be constructed using all
seven Tangram pieces. Explain why your list is complete.

(3.3) You made your seven piece Tangram by cutting up a *square*.
Which other quadrilaterals can be constructed using all seven
Tangram pieces? Make a complete list and prove that your list is
complete. (*If you get stuck do not look at the hints until you have had
a go at* (3.6)–(3.10).)

(3.4) In (3.3) you may have had considerable difficulty either trying
to construct, or trying to prove that it is impossible to construct a
quadrilateral with three angles of 45° and one angle of 225°. In the
traditional Tangram puzzle it is precisely the possibility of making
shapes with 'dents' which gives the puzzle its flexibility. But when you are
trying to find all possible shapes of a particular kind you do not want
too much flexibility! A polygon which has no corners pointing inwards
is called a *convex* polygon. (The polygon in Fig. 17.9(ii) is not convex.)

(3.5) Make a complete list of all convex polygons which can be
constructed using all seven Tangram pieces. Prove that your list is
complete. (First experiment and see how many you can find. Then try
to devise a systematic approach. *If you get stuck do not look at the
hints until you have had a go at* (3.6)–(3.10).)

(3.6) The problems in (3.3) and (3.5) are rather hard in spite of the
fact that it is easy enough to understand what you are supposed to do,
and in spite of the fact that the seven Tangram pieces provide plenty
of scope for experimenting. You presumably began by trying to make
as many shapes of the required kind as you possibly could. In (3.2)
you should have eventually realized firstly that the only possible
triangle would have to have two 45° angles and one 90° angle, and
secondly that there is only one such triangle with the same area as the

original Tangram square (Fig. 17.9(i)). The problem then comes down to deciding whether or not such a triangle can in fact be constructed using all seven Tangram pieces. This is not too difficult. But in (3.3) and (3.5) it is not at all easy to make the step from an initial exploration of the problem to a satisfactorily complete solution.

There are two difficulties here. The first is to find a simple systematic approach which offers some hope of finding all shapes of the required type. The second is to be absolutely sure that you have not missed any out.

Whenever you come up against a barrier like this it is worth looking at simpler problems of the same kind in the hope that ideas which work in the simpler problem may help you solve the original problem. Looking at a simpler problem also has a psychological advantage in that if you manage to complete the simpler problem successfully, this often gives you the confidence to go back and have another go at the harder problem.

(**3.7**) Draw an equilateral triangle with edge 10 cm long on a stiff card, mark it as in Fig. 17.10, and cut out your own seven piece 'Pangram'.

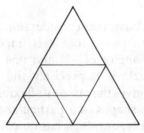

Fig. 17.10

(**3.8**) Find all possible triangles which can be constructed using all seven Pangram pieces. Explain why your list is complete.

(**3.9**) Find all possible quadrilaterals which can be constructed using all seven Pangram pieces. Prove that your list is complete.

(**3.10**) Find all convex polygons which can be constructed using all seven Pangram pieces. Prove that your list is complete.

When you have completed (3.8), (3.9), and (3.10) (consulting the hints if you get really stuck), go back and have another go at the problems in (3.3) and (3.5).

Hints

(3.2) What angles can one construct by fitting Tangram pieces together? So what can you say about the three angles of a triangle made with Tangram pieces? How many triangles have precisely these angles and also the same *area* as the original Tangram square (Fig. 17.9(i))? Now try to construct such a triangle.

(3.3) What are the possibilities for the four angles of a quadrilateral made with Tangram pieces? There are four different possibilities, and each possibility may correspond to zero, one, or several different quadrilaterals. Find all four possibilities before reading further.

The simple-minded approach is now to examine each of these four possibilities in turn.

(i) Suppose that all four angles are equal. You know one example of such a shape—namely the original square (Fig. 17.9(i)). You could start by looking for some others (for example, one other rectangle can be obtained by moving just one piece in the triangle you constructed in (3.2)). But at some stage you must make some kind of systematic search which will produce all possible rectangles without missing any out. What ideas could you use to simplify this systematic search? Have a go for yourself before reading further.

I shall outline one possible approach. Things will be a bit easier if you think of the original square as having a side of length 2 units, and area 4 square units. Then any rectangle you construct using all seven Tangram pieces will have to have area 4. This tells you what the product of the edge lengths must be. One way of exploiting this fact is to observe that each edge of the rectangle must be made up of edges of Tangram pieces, and that the edge lengths of Tangram pieces are rather special. *Work out the lengths of the edges of all seven Tangram pieces* (taking the edge length of the original square to be 2). This tells you what sort of lengths you can get by combining edges of Tangram pieces. *Find all possible pairs of such numbers having product equal to 4.* You should find exactly four possible pairs. You must then find some way of deciding which of these pairs correspond to genuine Tangram rectangles. Do this before reading further.

When trying to decide which pairs correspond to genuine Tangram rectangles you will probably have discovered another useful idea: namely, the two big triangles always have to be fitted in somehow. This immediately excludes the possibility of a rectangle with height <1. It is in fact easier to start with this idea (and not to bother about edge lengths), and to consider in turn the two possible orientations of a big triangle. (Why are there only two?)

There is another useful strategy which you may discover for yourself when trying to construct a 1×4 rectangle. Suppose you are trying to construct a trapezium or rectangle of given height h—say $h = 1$. This then forces certain pieces to be fitted in with a specific orientation (for example, the square has to fitted in as a diamond). Each shape whose orientation is determined in this way then contributes a specific amount to the cross-section at height $\frac{1}{2}h$. If the sum of these unavoidable contributions to the cross-section halfway up is too large, then no such shape can be constructed. (For example, for a 1×4 rectangle the two large triangles, the square, and the parallelogram each contribute 1 to the cross-section at height $\frac{1}{2}$. The middle-sized triangle then contributes an extra $\frac{1}{2}$—giving at least $4\frac{1}{2}$ altogether.)

(ii) By combining three of these four ideas (the fact that you know the total area; the fact that the large triangles have to be fitted in; and the fact that the height or other dimensions of the shape often determine the way certain pieces have to be fitted in, which then allows you to estimate other lengths such as the cross-section halfway up), you should be able to exclude or construct all the apparently possible quadrilaterals for each of the *two* sets of angles corresponding to trapezia, and the *one* set of angles corresponding to a parallelogram.

(iii) But it is not quite so clear how to exploit the second and third of these three ideas to exclude, or construct, all possible quadrilaterals with three angles of 45° and one angle of 225°. And if we want to use the other idea (that only certain special lengths can be obtained by sticking edges of Tangram pieces together) then we need to be a bit careful: because the two edges meeting in the 225° angle may not be made up of *whole edges* of Tangram pieces. (Why not?) However the two other edges must be constructed in this way. So a simple-minded approach might begin by combining the fact that we know the total area of the shape, with our earlier observation that the only lengths one can get by sticking edges of Tangram pieces together are lengths of the form $m + n(1/\sqrt{2})$, where m and n are whole numbers. The geometry of such a quadrilateral gives a simple formula for the area in terms of the lengths of these two sides, and the fact that m and n must be whole numbers can then be used.

(3.5) (i) A convex polygon cannot have any corners pointing inwards. So each of its angles must be <180°. There are precisely three angles <180° which can be constructed using Tangram pieces. What are they?

(ii) You know a formula for the sum of the angles in a polygon with n sides (Exercise 34, Chapter 9). So you know that the 'average angle' in a polygon with n sides is $(1/n)^{\text{th}}$ of this total angle-sum. You also know that none of the angles in a convex polygon made from Tangram pieces can be larger than the largest angle you found in (i). If a convex polygon constructed from Tangram pieces has n sides, find all possible values of n.

(iii) Now consider each value of n in turn. You have already found all convex polygons with 3 and with 4 sides in (3.2) and (3.3). It is not hard to show that the largest apparently possible value of n, which you found in (ii) above, cannot in fact occur (since almost all of the n corners would then need a piece with a 90° angle, and there simply aren't enough such pieces to go round). It is harder to deal with the next largest apparently possible value of n. To do this convincingly, and to check the two remaining values of n, you will have to devise some clever systematic method which will allow you to check every single possibility and to know for sure that you have not missed any out. But start by experimenting with each value of n. (You should not just be looking to see which convex polygons you can and which you cannot construct, but you should be trying to get some insight into what it is that makes some convex polygons constructible and others not. You will need this insight when you come to devise a systematic way of finding all possible convex polygons which does not miss any.)

(**3.8**) What are the possibilities for the three angles in a Pangram triangle? How many triangles are there with these three angles and having the same area as the original Pangram triangle (Fig. 17.10)?

(**3.9**) (i) What angles can one construct by fitting Pangram pieces together? What are the possibilities for the four angles of a Pangram quadrilateral?

(ii) If a quadrilateral has two large (equal) angles and two small (equal) angles, then the two large angles may be either adjacent to one another or opposite to one another. You must now consider each of these possible arrangements in turn.

(iii) Knowing the angles tells you the kind of quadrilateral you are looking for, but it does not tell you the exact shape. You know what the area of the quadrilateral must be. But what can you say about the lengths of its edges, its height, its width, and so on? One very useful idea, no matter which shape you are trying to construct, is that the large triangles have to be fitted in somehow. This gives you an absolute minimum for the 'height' of the shape measured from the base of one of the large triangles. Another very useful idea is the fact that the edges of the Pangram pieces are all simple multiples of the edge of the smallest triangle. So if you call the edge of one of these small triangles 1 unit, then the edge lengths of any quadrilateral you can construct must be a whole number of units. An even more useful idea is the fact that each of the seven Pangram pieces is actually made up of a whole number of these small triangles—which we shall call 'mini-units'. Hence every shape you can construct consists of exactly 16 of these mini-units. If you use these simple ideas carefully, you should be able to devise a systematic approach which will not only produce all possible Pangram quadrilaterals, but which will at the same time prove that you have not missed any out.

(**3.10**) (i) A convex polygon cannot have any corners pointing inwards, so each of its angles has to be $<180°$. There are only two angles $<180°$ which can be constructed using Pangram pieces. What are they?

(ii) You know a formula for the sum of the angles in a polygon with n sides (Exercise 34, Chapter 9). So you know that the average size of an angle in a polygon with n sides must be $(1/n)^{th}$ of this total angle-sum. You also know that none of the angles in a convex polygon constructed from Pangram pieces can be larger than the largest angle you found in (i). You should therefore be able to list all values of n for which there could conceivably exist a convex polygon with n sides which uses all seven Pangram pieces. What are these possible values of n?

(iii) Now consider each value of n in turn. You have already found all convex polygons with 3 and with 4 sides in (3.8) and (3.9). Try to find all convex polygons with 5 sides and with 6 sides before reading on.

Suppose it were possible to construct a convex pentagon using all seven Pangram pieces. What would its angles have to be? Once you know the angles you can sketch the shape and see that it must be obtained by chopping off an equilateral triangle from the corner of a parallelogram. What are the possibilities for the area A of this missing equilateral triangle (measured in mini-units)? Which parallelograms can one construct from $16 + A$ mini-units?

Finally suppose that it were possible to construct a convex hexagon. What would its angles have to be? What does this tell you about the three pairs of opposite sides of the hexagon? Show that no side can be more than 2 units long. How many sides can have length 2 units? How many sides must have length 1 unit? Use this to construct all possible convex hexagons which use all seven Pangram pieces.

---

4. The traditional Towers of Hanoi puzzle has three pegs, with a tower of rings of decreasing size on one of the pegs. The challenge is to find the least number of moves required to transfer the tower onto another peg, moving one ring at a time and never placing a larger ring on top of a smaller ring. What is the least number of moves if there are more than three pegs?

Outline

(4.1) Start by making sure that you know the least number of moves required in the three peg problem, and that you can prove that this number is the smallest possible number of moves. [Make yourself a tower of rings of decreasing sizes, or of pieces of paper (A4, A5, A6, ...), or of coins (50p, 10p, 2p, 5p, 20p, 1p), and imagine three pegs or squares (left, right, and centre). Then experiment with towers of different sizes until you think you know the least number of moves required to transfer the tower from its initial peg or square to a different peg or square—never placing a larger ring on top of a smaller one. Fill in the table in Fig. 17.11 at least as far as $r = 7$.

Number of rings $= r$	0	1	2	3	4	5	6	7	8
Least number of moves $= m$									

Fig. 17.11

Find a formula for the least number of moves m in terms of the number of rings r. Explain why your formula really does give the least number of moves.]

(4.2) Now experiment with *four* pegs or squares, and fill in the table in Fig. 17.12 at least as far as $r = 7$.

Number of rings $= r$	0	1	2	3	4	5	6	7	8
Least number of moves *you* can achieve									

Fig. 17.12

(4.3) Trial and error is obviously not a good enough method. The four-peg problem is more complicated than the three-peg problem, and it is almost impossible to be sure that the 'least number of moves *you* can achieve' is in fact the smallest number of moves required. So you are more or less forced to resort to the 'guess–check–revise your guess–check again– . . . –prove' procedure. You should probably *not* start by *guessing* the least number of moves required for 8 rings on the basis of the entries you have just made in the above table—after all, we have already remarked that these entries are probably not entirely accurate beyond $r = 5$ or $r = 6$. Instead it might be better to begin by looking for *a general strategy for shifting any number of rings from one peg to another* which gives at least as good results as your entries in the table of Fig. 17.12 (and possibly better). Do this now.

(4.4) It is not easy to check your general strategy in any simple-minded way. Suppose, for example, that your strategy predicts that the least number of moves required to transfer 8 rings is 33. It may even tell you how to actually carry out the transfer in this number of moves. How can you be absolutely sure that it is not somehow possible—by being really clever—to do it in 31 moves? (The least number of moves required is bound to be odd. Can you see why?) Checking every possible way of transferring 8 rings is out of the question. So how can you begin?

The first thing you might try to do is to tidy up your general strategy until it looks convincing. Next you might ask whether the basic steps which your general strategy requires are the kind of moves one would expect to have to make in order to transfer a tower of r rings as efficiently as possible.

(4.5) Another thing which might be sensible is to compare the least number of moves your strategy predicts for the four-peg problem against what you know (from (4.1)) to be the least number of moves in the three-peg problem. If you can see some connection between the pattern of numbers in the three-peg problem and the pattern of numbers your strategy predicts for the four-peg problem, then you might begin to feel that you are on the right lines. You may then have enough confidence in your guess to try to prove that it is correct. So fill in the table in Fig. 17.13 at least as far as $r = 20$ and see if you can spot some connection between the pattern of numbers in the three-peg problem and the pattern of numbers your strategy predicts for the four-peg problem.

Number of rings $= r$	Least number of moves required in the 3-peg problem $= M_3(r)$	Number of moves your strategy requires in the 4-peg problem
0	0	0
1	1	1
2	3	3
3	7	
4	15	
5	31	
6		

Fig. 17.13

Number of rings $= r$	Least moves in 3-peg problem	Predicted minimum in 4-peg problem	Predicted minimum in 5-peg problem
0	0	0	0
1	1	1	1
2	3	3	3
3	7	5	
4	15	9	
.			
.			
.			
16			

Fig. 17.14

(4.6) Your general strategy for a tower of r rings in the four-peg problem may make use of the fact that you know all about the three-peg problem, and the four-peg problem *for fewer than r rings*. If so, then it should not be hard to generalize your strategy to predict the least number of moves for transferring r rings in the *five*-peg problem (Fig. 17.14). You would obviously like the pattern of predictions for the five-peg problem to fit in with the patterns you observed in (4.5).

(4.7) Once you have arrived at a strategy for the four-peg problem which predicts what you believe to be the absolute minimum number of moves required to transfer r rings from one peg to another, try to show that it is impossible to improve on your strategy.

Hints

(4.1) You should have no trouble finding the correct value of m when $r = 0, 1, 2, 3$, or 4. Try to get the correct value for $r = 5$. If you cannot see a pattern, fill in the slightly different table in Fig. 17.15.

Number of rings $= r$		0	1	2	3	4	5
(Least number of moves) $+ 1 = m + 1$		1	2				

Fig. 17.15

Guess a formula for $m + 1$, then try to explain why it is always valid. (Suppose you want to transfer r rings from peg A onto peg B. At some stage you will have to shift the largest ring from peg A onto peg B. Where must the other $r - 1$ rings be when you make this move? How many moves did it take to get them there?)

(4.3) What could your 'general strategy for transferring r rings from one peg to another in the least number of moves' possibly depend on? You certainly know the least number of moves required to transfer any number of rings in the *three*-peg problem. And a strategy for r rings in the *four*-peg problem can presumably make use of the fact that you have already discovered how to solve the *four*-peg problem for *fewer than r rings*.

(4.5) One very natural thing to do when trying to make sense of a sequence of numbers, or to compare two sequences, is to look at the differences between successive terms (see Fig. 17.16).

(4.6) Read the hint to (4.5).

(4.7) It should be clear by now that the process of transferring $r + 1$ rings from one peg to another always has the same basic structure: (1) transfer the top r rings from the original peg, leaving one peg completely free; (2)

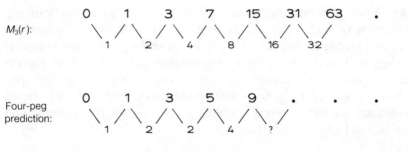

Fig. 17.16

transfer the largest ring onto the free peg; (3) transfer the other r rings back on top of the largest ring. The least number of moves required for steps (1) and (3) are closely related! Can you see how, and why? Once you realize this, all you need to do is to find the most efficient way of completing step (1).

Thus the key to the solution of our main problem of transferring $r + 1$ rings using p pegs lies in solving the *intermediate* problem of 'finding the least number of moves required to transfer the top r rings onto $p - 2$ pegs (leaving one peg free and one peg occupied by the largest ring only)'. What makes the *three*-peg problem ($p = 3$) so appealing (and potentially misleading) is that in that case we have $p - 2 = 1$; thus the 'intermediate problem' that we need to solve in order to transfer $r + 1$ rings as efficiently as possible is *exactly the same problem*, but with r rings instead of $r + 1$. However, when the number of pegs is four or more, the intermediate problem which holds the key to the solution of our original problem is rather different. So perhaps you should abandon the original problem for a while, and concentrate on solving the intermediate problem first: 'given r rings on one of p pegs, find the least number of moves required to transfer these r rings onto $p - 2$ other pegs leaving two pegs free'.

> 5. (*a*) What loci do the familiar focus-directrix definitions of conics give in Taxicab geometry? What loci do the familiar focus-focus definitions of conics give in Taxicab geometry? Are the two families related? (If so, how? If not, why not?)
> (*b*) Are these loci related in some way to the plane cross-sections of a Taxicab circular cone?

Outline

(**5.1**) Perhaps the first thing to do is to refresh your memory as to how the 'focus-directrix' and 'focus-focus' definitions work in ordinary

Euclidean geometry, and why they give rise to the same loci. If all this is second nature to you, then go straight on to (5.2). If on the other hand you know nothing at all about conics in ordinary Euclidean geometry, then you should work carefully through the following sequence of exercises—if necessary in association with a text on the geometry and analytical geometry of conics. At several points you will find it useful to have read *Introduction to geometry* by H. S. M. Coxeter (Wiley, 1969, pp. 115–19), and especially *What is mathematics?* by R. Courant and H. Robbins (Oxford University Press, 1961, pp. 198–201).

Focus-directrix (i) Draw a straight line *l* and mark a point *P* not on *l*.

(*a*) Sketch the locus of all points which are equidistant from both *l* and *P*. This locus is called a *parabola: P* is its *focus,* and *l* its *directrix.*

(*b*) The given line *l* and the point *P* suggest two natural axes, namely *l* itself (say as *y*-axis) and the line through *P* perpendicular to *l* (as *x*-axis) cutting *l* at *O* (the origin). The point halfway between *O* and *P* lies on the parabola, so it is natural to think of the distance *OP* as being equal to 2*a* for some *a* (so *P* has coordinates (2*a*, 0)). Find the equation of the locus.

(*c*) A doubly-infinite, hollow, circular cone, with apex at the origin is obtained by rotating a straight line through the origin about the (vertical) *z*-axis. Any straight line which lies on the cone is called a 'generator'. Cut the cone by a plane parallel to a generator. This gives rise to a cross-sectional curve *γ* which is in one piece but which has two ends going off to infinity. Imagine a sphere inserted inside the cone between the cutting plane and the apex. This sphere is then pumped up until it fits tightly inside the cone and touches the cutting plane in a point *P*. The sphere touches the cone along a circle *c*. The plane through *c* meets the original cutting plane in a line *l*. Show that the original curve *γ* is a parabola with focus *P* and directrix *l*. (See R. Courant and H. Robbins *What is mathematics?*, pp. 199–201. Oxford University Press, 1961.)

(ii) Draw a straight line *l* in the plane and mark a point *P* not on *l*.

(a) Sketch the locus of all points which are half as far from *P* as they are from *l*. This locus is called an *ellipse* with eccentricity $e = \frac{1}{2}$: *P* is a *focus,* and *l* a *directrix.* If $0 < e < 1$, then the locus of all points whose distance from *P* is *e* times their distance from *l* is an ellipse with eccentricity *e*.

(b) The given line *l* and the point *P* suggest two natural axes, namely *l* itself (say as *y*-axis) and the line through *P* perpendicular to *l* (as *x*-axis) cutting *l* at *O* (the origin). The point dividing *OP* in the ratio 1 : *e* lies on the ellipse, so it is natural to think of the distance *OP* as being equal to $c(1 + e)$ for some *c* (so *P* has coordinates

$(c(1 + e), 0))$. Find the equation of the locus. Then rewrite your equation in terms of the two variables $(x - c/(1 - e))$ and y. (The fact that the equation involves only the squares of these two quantities shows that the ellipse has reflection symmetry in the x-axis and in the line $x = c/(1 - e)$, so that there is a second focus at $(c(1 + e^2)/(1 - e), 0)$, and a second directrix at $x = 2c/(1 - e)$.)

(c) Cut an (infinite, hollow, circular) cone by a plane to obtain a cross-section which is a closed curve. Explain why this cross-section is an ellipse. Find its eccentricity (in terms of the half-angle of the cone and the inclination of the cutting plane).

(iii) Draw a straight line l and mark a point P not on l.

(a) Sketch the locus of all points which are twice as far from P as they are from l. This locus is called a *hyperbola* with eccentricity $e = 2$: P is a *focus*, and l a *directrix*. If $e > 1$, then the locus of all points whose distance from P is e times their distance from l is an ellipse with eccentricity e.

(b) The given line l and point P suggest two natural axes, namely l itself (as y-axis) and the line through P perpendicular to l (as x-axis) cutting l at O (the origin). The point dividing OP in the ratio $1 : e$ lies on the hyperbola, so it is natural to think of the distance OP as being equal to $c(1 + e)$ for some c (so P has coordinates $(c(1 + e), 0)$). Find the equation of the locus. Then rewrite your equation in terms of the two variables $(x + c/(e - 1))$ and y. (The fact that the equation involves only the squares of these two quantities shows that the hyperbola has reflection symmetry not only in the x-axis, but also in the line $x = c/(1 - e)$, so there is a second focus at $(c(1 + e^2)/(1 - e), 0)$, and a second directrix at $x = 2c/(1 - e)$.)

(c) Cut a (doubly-infinite, hollow, circular) cone by a plane which intersects both the downward pointing part of the cone and the upward pointing part. The cross-section then has two branches. Explain why this cross-section is a hyperbola. Find its eccentricity (in terms of the half-angle of the cone and the inclination of the cutting plane).

Remarks: You will have noticed that parts (ii) and (iii) above are absolutely identical except for the fact that in (ii) we assume that $0 < e < 1$, and in (iii) we assume that $e > 1$. Part (i) is slightly different (partly because, when $e = 1$, all those expressions with $e - 1$ in the denominator become meaningless).

(a) But suppose we fix the directrix l and the focus P and consider ellipses with eccentricity e getting closer and closer to 1. Then the other focus has coordinates $(c(1 + e^2)/(1 - e), 0)$, and while the numerator $c(1 + e^2)$ is always $> c$, the denominator $1 - e$ remains > 0

but gets closer and closer to 0. So the other focus moves quickly off along the x-axis towards $+\infty$ (as does the other directrix with equation $x = 2c/(1 - e)$). Thus if l and P are fixed, then as e tends to 1 we get a whole family of ellipses which tend towards a curve with only one (finite) focus P, and only one (finite) directrix l: in other words, the parabola with focus P and directrix l.

(b) Similarly if we fix the directrix l and the focus P and consider hyperbolas with eccentricity e getting closer and closer to 1, then the other focus has coordinates $(c(1 + e^2)/(1 - e), 0)$, and so moves quickly off along the negative x-axis towards $-\infty$ (as does the other directrix with equation $x = 2c/(1 - e)$, and the second branch of the hyperbola which lies beyond this second directrix). Thus if l and P are fixed, then as e tends to 1 we get a whole family of hyperbolas which tend towards a curve with only one branch, one finite focus P, and one directrix l: in other words, the parabola with focus P and directrix l.

(c) Think of a plane which cuts a doubly-infinite, hollow, circular cone. When the plane cuts the axis of the cone at right angles, the cross-section is a circle. As the plane tilts, the cross-section becomes an ellipse. And as the plane tilts more and more, the ellipse grows longer and longer, until eventually the plane cuts the cone parallel to a generator. The cross-section then becomes a parabola. If the plane continues to tilt, it begins to cut the upward pointing section of the doubly-infinite cone and we get a curve with two branches, namely a hyperbola.

Remarks (a), (b), and (c) suggest that one should think of the *parabola* (with $e = 1$, and cutting plane inclined to the axis of the cone at the same angle as a generator of the cone) as a curve which is intermediate between *ellipses* (with $e < 1$, and cutting plane inclined to the axis of the cone at a larger angle than a generator of the cone) and *hyperbolas* (with $e > 1$, and cutting plane inclined to the axis of the cone at a smaller angle than a generator of the cone).

Focus-focus: (ii) (a) Let P and Q be two focuses of the ellipse whose equation you found in (ii) above. If X is any point on the ellipse, show that the sum of its distances from P and from Q, namely $|PX| + |QX|$, is always equal to $2ec/(1 - e)$.

(b) Given any two points P', Q' in the plane, and a number $d > |P'Q'|$, sketch the locus of all points X' for which $|P'X'| + |Q'X'| = d$. Explain why this locus is an ellipse, and why P', Q' must be its two focuses. (See R. Courant and H. Robbins, *What is mathematics?*, pp. 199–201. Oxford University Press, 1961.) Find its eccentricity e, and the distance $c(1 + e)$ between each focus and its associated directrix.

(iii) (a) Let P and Q be the two focuses of the hyperbola whose equation you found in (iii) above. If X is any point on the hyperbola, show that the absolute value of the difference of its distances from P and from Q, namely $||PX| - |QX||$, is always equal to $2ec/(e-1)$.

(b) Given any two points P', Q' in the plane and a number $d < |P'Q'|$, sketch the locus of all points X' for which $||P'X'| - |Q'X'|| = d$. Explain why this locus is a hyperbola, and why P', Q' must be its two focuses. Find its eccentricity e and the distance $c(1 + e)$ between each focus and its associated directrix.

(5.2) In the rest of this investigation you may find it helpful to work on square dotty paper as in Fig. 17.17. Choose any dot P and sketch the (Taxicab) circles with centre P and radius 1, 2, 3, 4.

Fig. 17.17

(5.3) *Focus-directrix:* In Euclidean geometry it does not matter which line you choose as directrix; neither does it matter which point you choose as focus. The only things that affect the shape of the locus are the eccentricity e, and the distance $c(1 + e)$ between the focus and its directrix. In Taxicab geometry it still does not matter which point you choose as focus; but the definition of distance in Taxicab geometry ('along-and-up', or 'up-and-along', or . . .) singles out lines parallel to the x-axis and lines parallel to the y-axis as rather special. So you might expect the orientation of the directrix to affect the shape of a locus with the 'focus-directrix' property.

(i) Choose any point P as focus, and a directrix l which is 'vertical'—say with equation $x = -6$ relative to P as origin.

(a) Sketch the two 'ellipses' with focus P and directrix l having eccentricities $e = \frac{1}{2}$ and $e = \frac{1}{5}$.

(b) Sketch the 'parabola' with focus P and directrix l (having eccentricity $e = 1$).

(c) Sketch the 'hyperbolas' with focus P and directrix l having eccentricities $e = 2$ and $e = 5$. (You should expect each 'hyperbola' to have *two* branches.)

(ii) Choose any point P as focus, and a directrix l which has slope 2—say with equation $y = 2x + 12$ relative to P as origin.

(*a*) Sketch the 'ellipses' with focus P and directrix l having eccentricities $e = \frac{1}{2}$ and $e = \frac{1}{5}$.

(*b*) Sketch the 'parabola' with focus P and directrix l (having eccentricity $e = 1$).

(*c*) Sketch the 'hyperbolas' with focus P and directrix l having eccentricities $e = 2$ and $e = 5$.

(iii) Choose any point P as focus, and a directrix l which has slope 1—say with equation $y = x + 6$ relative to P as origin.

(*a*) Sketch the 'ellipses' with focus P and directrix l having eccentricities $e = \frac{1}{2}$ and $e = \frac{1}{5}$.

(*b*) Sketch the 'parabola' with focus P and directrix l (having eccentricity $e = 1$).

(*c*) Sketch the 'hyperbolas' with focus P and directrix l having eccentricities $e = 2$ and $e = 5$.

(iv) In what sense do the three orientations of the directrices in Parts (i), (ii), and (iii) represent all possible cases?

(v) The 'ellipse' with focus P and directrix l in (i)(*a*) which has eccentricity $e = \frac{1}{2}$ passes through the point $(-2, 0)$. Consider all 'ellipses' which have P as focus which have a vertical directrix, and which pass through the point $(-2, 0)$ (relative to P as origin). When l has equation $x = -n$, such an ellipse has eccentricity $e = 2/(n - 2)$. What happens to these ellipses as $n \to \infty$? (This shows that the focus-directrix definition of an 'ellipse' includes the 'circles' of (5.2) only as a limiting case.)

(5.4) *Focus-focus:* In (5.3) you saw how different orientations of the directrix give rise to different looking 'conics' in Taxicab geometry. In the same way, you might expect the shape of a locus with one of the 'focus-focus' properties (either $|PX| + |QX| =$ constant, or $||PX| - |QX|| =$ constant) to depend on the orientation of the line PQ joining the two 'focuses'. Choose any point P as the first 'focus'.

(i) Choose the second 'focus' Q to have coordinates $(4, 0)$ relative to P as origin.

(*a*) Sketch the locus of all points X satisfying $|PX| + |QX| = 6$, the locus of all points X satisfying $|PX| + |QX| = 7$, and the locus of all points X satisfying $|PX| + |QX| = 8$. Are these 'focus-focus' curves like any of the 'ellipses' you found using the 'focus-directrix' property in (5.3)?

(*b*) Sketch the locus of all points X satisfying $||PX| - |QX|| = 3$, the locus of all points X satisfying $||PX| - |QX|| = 2$, and the locus of all points X satisfying $||PX| - |QX|| = 1$. Are these 'focus-focus' curves like any of the 'hyperbolas' you found using the 'focus-directrix' property in (5.3)?

(ii) Choose the second 'focus' Q to have coordinates $(3, 1)$ relative to P as origin.

(*a*) Sketch the locus of all points X satisfying $|PX| + |QX| = 6$, the locus of all points X satisfying $|PX| + |QX| = 7$, and the locus of all points X satisfying $|PX| + |QX| = 8$. Are these 'focus-focus' curves like any of the 'ellipses' you found using the 'focus-directrix' property in (5.3)?

(*b*) Sketch the locus of all points X satisfying $||PX| - |QX|| = 3$, the locus of all points X satisfying $||PX| - |QX|| = 2$, and the locus of all points X satisfying $||PX| - |QX|| = 1$. Are these 'focus-focus' curves like any of the 'hyperbolas' you found in (5.3) using the 'focus-directrix' property?

(iii) Choose the second 'focus' Q to have coordinates $(2, 2)$ relative to P as origin. Sketch the locus of all points X satisfying $|PX| + |QX| = 6$, and the locus of all points X satisfying $||PX| - |QX|| = 2$. Are these 'focus-focus' curves like any of the 'ellipses' or 'hyperbolas' you found in (5.3) using the 'focus-directrix' property?

(iv) Choose the second 'focus' Q to be equal to P. Sketch the locus of all points X satisfying $|PX| + |QX| = 4$. Compare this 'focus-focus' curve with the curves you found in (5.2) and with the limit curve you found as $n \to \infty$ in (5.3)(v).

(v) Parts (i)–(iii) suggest that the 'focus-focus' property does not fit in any obvious way with the 'focus-directrix' Taxicab conics of (5.3). But Part (iv) shows that, when the two focuses are equal, the 'focus-focus' property matches up exactly with the familiar definition of a (Taxicab) circle. Can you see a sense in which all the 'focus-focus' curves are simply generalizations of a circle?

(5.5) So far as one can tell at this stage, there is no apparent connection in Taxicab geometry between curves with the 'focus-directrix' property and curves with the 'focus-focus' property. Moreover, if you understood what (5.4)(v) was getting at, then you will agree that curves with the 'focus-focus' property look like rather uninteresting generalizations of the circle. In Euclidean geometry curves with the 'focus-directrix' property (or the 'focus-focus' property) can also be obtained as plane cross-sections of a doubly-infinite right circular cone. Is there any connection in Taxicab geometry between curves with either the 'focus-directrix' property, or the 'focus-focus' property, and plane cross-sections of a doubly infinite (Taxicab) circular cone?

(i) What is a circle in Taxicab geometry? So which familiar object plays the role of a 'circular cone' in Taxicab geometry?

(ii) In Euclidean geometry the only thing that affects the shape of a plane cross-section of a given circular cone is *the inclination of the cutting plane.* (Moving the cutting plane up and down while keeping the same inclination to the vertical affects the *size* of the cross-sectional curve, but not its *shape.* Similarly rotating the cutting plane about the vertical axis of the cone has no effect on the cross-section because of the cone's symmetry.) In Taxicab geometry a 'circular' cone does not have the same symmetry about its axis: rotating the cutting plane about the vertical axis of the cone gives rise to *different* shaped cross-sectional curves.

- (a) Call a plane cross-section of a doubly-infinite (Taxicab) circular cone *'elliptical'* if the cutting plane cuts right across one of the infinitely extended portions of the cone giving rise to a cross-section which is a closed curve. Sketch the different kinds of elliptical cross-sections one gets as a cutting plane is rotated about the axis of the cone. Do these bear any resemblance to the curves you found in (5.3)(i)(*a*), (5.3)(ii)(*a*), (5.3)(iii)(*a*), or to the curves you found in (5.4)(i)(*a*), (5.4)(ii)(*a*), (5.4)(iii)?

(b) Now do the same for 'parabolic' cross-sections of a (Taxicab) circular cone. Do the curves you get bear any resemblance to the curves you found in (5.3)(i)(*b*), (5.3)(ii)(*b*), (5.3)(iii)(*b*), or to the curves you found in (5.4)(i)(*b*), (5.4)(ii)(*b*), (5.4)(iii)?

(c) Now do the same for 'hyperbolic' cross-sections.

(5.6) If you sketched the cross-sections in (5.5) accurately, then you will have recognized that though their general shape corresponds to curves with the 'focus-directrix' property, they do not match up exactly.

(i) For example, one can never get a cross-section of a Taxicab circular cone which looks exactly like the curve you found in (5.3)(iii)(*b*). Why not?

(ii) What is it about the other curves with the 'focus-directrix' property which makes them slightly different from the cross-sections of a cone?

(5.7) In spite of their differences, the similarity between the curves with the 'focus-directrix' property which you found in (5.3) and the cross-sections of a (Taxicab) circular cone which you found in (5.5) is sufficiently striking to warrant further investigation. Here are a few possible starting points.

(i) Suppose you try to identify the 'focus' and the 'directrix' of the conic associated with a cross-section of a (Taxicab) circular cone using the method of Dandelin's spheres (see R. Courant and H. Robbins, *What is mathematics?*, pp. 200–201. Oxford University Press, 1961). What is a sphere in Taxicab geometry? If you want to insert a sphere into a Taxicab circular cone in a sensible way, what does this suggest about the apex angle of the cone? Where will the sphere touch the cutting plane?

(ii) If in (5.5)(ii)(*b*) the cutting plane is parallel to one of the faces of the cone one obtains a 'parabolic' cross-section similar to the 'parabola' in (5.3)(iii)(*b*), with the two 'angles' equal but considerably less than the 135° angle in (5.3)(iii)(*b*).

(a) What natural transformation of the 'parabolic' cross-section would produce a 'parabola' in the sense of (5.3)(iii)(*b*)?

(b) Where do the two infinite arms of the 'parabola' in (5.3)(iii)(*b*) meet? Where do the two infinite arms of the 'parabolic' cross-section described above meet?

(iii) Where do opposite sides of the 'ellipses' in (5.3) meet? Where do opposite sides of the 'elliptical' cross-sections in (5.5)(i) meet?

Hints

(**5.2**) The (Taxicab) circle of radius 1 must go through the four points *N*, *S*, *E*, *W* which are respectively one unit above, below, to the right, and to the left of *P*. But it must also go through the point *X* which is half a unit to the right and half a unit above *P* (since $\frac{1}{2} + \frac{1}{2} = 1$).

(**5.3**) Each of the 'curves' you are asked to sketch is made up of *straight line segments*. Watch out for the points where the slope suddenly changes! How are these 'corners' related to the focus and directrix?

(**5.5**) (i) See (5.2).

(ii) The curves are not *exactly* the same as those you have met before; but they are clearly related to those curves in some way (see (5.6)).

(**5.7**) (i) You would like the (Taxicab) sphere to fit 'snugly' inside the apex of the (Taxicab) circular cone.

(ii) (*a*) Project the 'parabolic' cross-section onto the horizontal plane through the apex of the cone.

6. Is there a general pattern governing flips in
any base?

Outline

Though you know all about flips in base 10 from Chapter 7 and
Chapter 11, you know rather less about what happens in other bases.
You discovered in Chapter 8 that the flips you found in base 10 are
part of a general pattern of flips in any base (see the solution to
Exercise 25(iv) at the end of Chapter 7). But unless you pursued the
Project at the end of Chapter 8 fairly seriously, you probably did not
enquire too closely whether this general pattern accounted for *all* flips
in any base. So you will probably have to start by making *complete*
lists of flips in base b for small values of b. To keep things simple it
might be a good idea to begin by concentrating on flips with up to
four digits. You should extend what you find to cover flips with more
than four digits later, by which time you should have a clearer idea
what you expect to find.

(**6.1**) Suppose we ignore examples like $01 \times 2 = 10_{\text{base } 2}$ (at least to
start with). Find all flips in base 2.

(**6.2**) (i) Find all flips with two digits in base 3.
 (ii) Find all flips with three digits in base 3.
 (iii) Find all flips with four digits in base 3.

(**6.3**) Repeat (6.2) for flips in base 4.

(**6.4**) Repeat (6.2) for flips in base b for $b = 5, 6, 7, 8, 9$.

(**6.5**) (i) What flips do you expect to find in base 11?
 (ii) Repeat (6.2) for flips in base 11.

(**6.6**) You should by now have enough to think about. It may be a
good idea to begin by sorting out exactly what happens for flips with
two, three, and four digits in *all* bases before going on to consider
flips with more than four digits. Ideally, given the base (say base b),
you would like to have a recipe for constructing all flips with n digits
in base b. You would also like to know how many flips with n digits to
expect. As you try to make sense of the evidence two other questions

should strike you—questions which we noticed back in Chapter 8: the first concerns the connection between flips in one base and flips in another (see, for example, the solution to Exercise 25(iv) at the end of Chapter 7); the second concerns 'x-flips, where x is not a whole number' (see the solution to Exercise 31 at the end of Chapter 8).

Hints

(6.1) It seems sensible to start by concentrating on 'multipliers' which are $<b$ (though later on you may want to interpret $01 \times 2 = 10_{\text{base }2}$ as the first instance of a general pattern).

(6.2) (i) For the moment it seems sensible to ignore palindromes — since every palindrome is a 1-flip in every base (though later on you may want to reinterpret $11 \times 1 = 11_{\text{base }3}$, for example, as a rather special palindrome).

(6.4) Since (6.1), (6.2), and (6.3) did not produce anything unexpected, and since you know exactly what happens in base 10, you may be tempted to try to cut corners here. Don't! Every one of these values of b has something to tell you.

(6.5) (i) Before guessing what will happen in base 11, try to see why the surprises you found in (6.4) do not occur when $b = 10$.

Don't bite my finger — look where it's pointing.
(Warren McCulloch, *Platform for change*)

Index